Antonio Marussi

Intrinsic Geodesy

Translated by W. I. Reilly

With 7 Figures

Springer-Verlag
Berlin Heidelberg New York Tokyo

Professor ANTONIO MARUSSI, Via C. Battisti 31, I-34 125 Trieste

Translator: Dr. W. IAN REILLY, Department of Scientific and Industrial Research, Geophysics Division, P.O.B. 1320, Wellington/New Zealand

ISBN 3-540-15133-8 Springer-Verlag Berlin Heidelberg New York Tokyo
ISBN 0-387-15133-8 Springer-Verlag New York Heidelberg Berlin Tokyo

Library of Congress Cataloging in Publication Data. Marussi, Antonio. Intrinsic geodesy. Bibliography: p. Includes indexes. 1. Geodesy. 2. Spaces, Generalized. 3. Coordinates. I. Title. QB283.M38 1985 526'.1 85-2737

Printed in Germany

Typesetting: K. u. V. Fotosatz Beerfelden
Offsetprinting, and binding: Konrad Triltsch, Graphischer Betrieb, Würzburg
2132/3130-543210

A te Lori che mi sei stata vicina fiduciosa nella carriera scientifica e che sei ora il mio solo sostegno.

Con molto affetto
Antonio

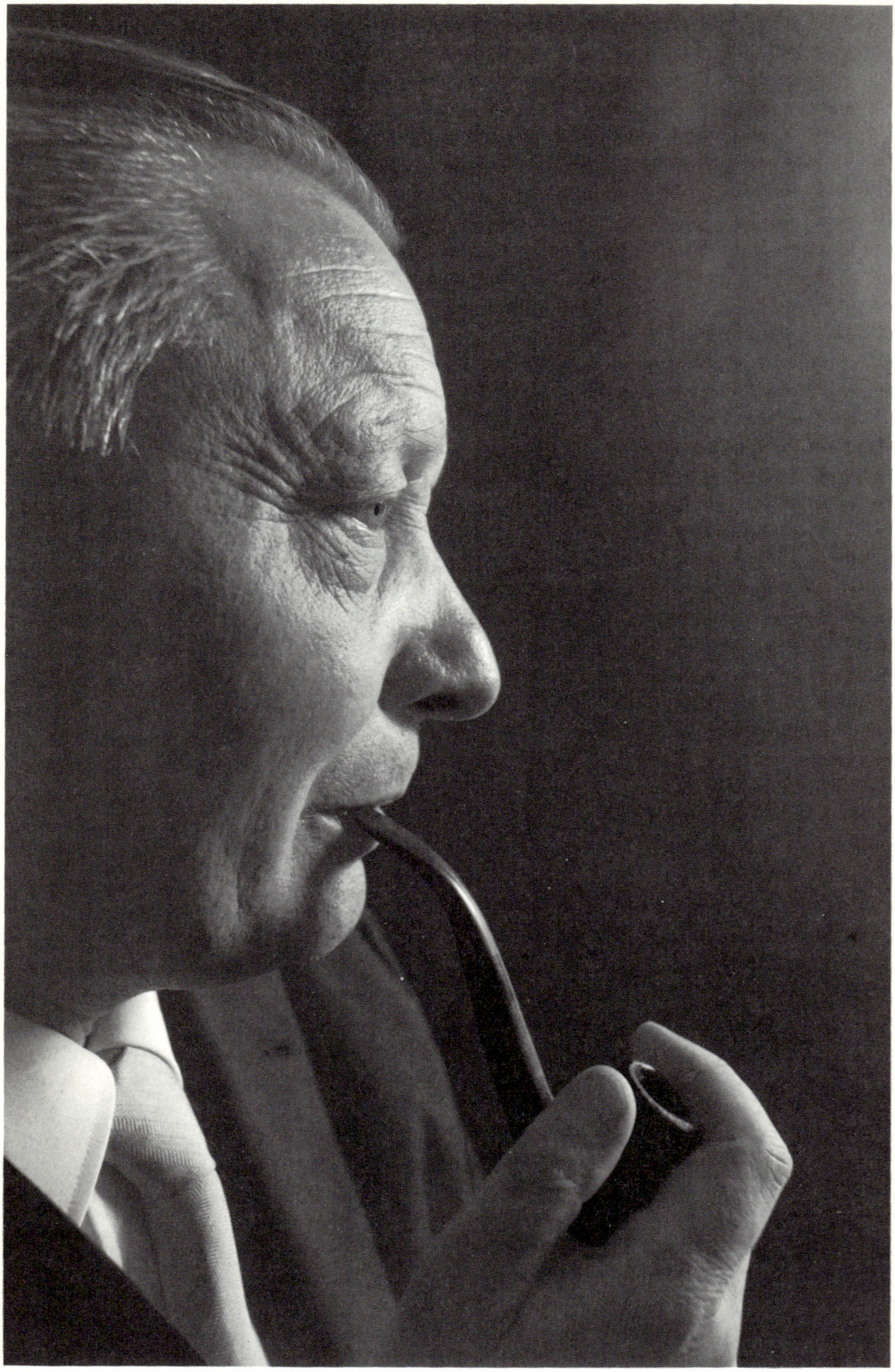

Antonio Marussi

Antonio Marussi bestrode the world like a colossus. From Chile to China, from the roots of the Himalayas to the orbits of artificial satellites, his questing mind probed deeply and ranged widely to understand the inwardness of the Earth. At once sceptical and imaginative, he was not content with the conventional view, and in his reformulation of three-dimensional geodesy, as in his explorations of the Himalayan regions, set off on different and more fundamental paths than his contemporaries had trodden. His geodetic theory required the agility of a mathematical scholar, but he was also a person who confronted Nature in practice and overcame her, whether on the peaks of the high Himalayan mountains or in making delicate apparatus in the laboratory. He was a forceful organiser, who took advantage of the special characteristics of the Grotta Gigante to set up the pendulums with which he saw the signatures of the spectrum of the Earth, detected earthquakes before they happened and found that the Moon does not shield us from the gravity of the Sun. Thinker, observer, organiser, all these he had to be to bring to fruition his studies of those high regions north of the Himalayas which we must know if we are to understand the tectonics of the continents, yet where still so few observations have been made. Some 25 years after his first expeditions to the Hindu Kush, he proposed the first long seismic line from the Pamirs to the plains of India. Thought, observations, organisation, all indeed were essential, but still more was called for, and still more he gave. Fortified by the high reputation he already enjoyed in the East, Antonio Marussi became diplomat, and brought the scientists of four nations to co-operate with him.

Antonio Marussi left us thinking of the gravity field of the Earth and of geodesy in ways very different from those he found. That was a major intellectual achievement, and its value became outstandingly clear when geodesy came to use artificial satellites. He established his own observatory on his beloved Carso and pursued collaboration with his neighbours of mountainous countries. His last work, like some of his earliest at Trieste, took him to the wastes of Asia and to the immense high plateaux that hide secrets of the origins of the great mountains he loved and conquered.

We lament indeed the untimely, unlooked-for passing of Antonio Marussi, a proud Triestino, whose appreciation of other countries was based firmly on his loyalty to that peculiar city, but we shall treasure with gratitude these works with which he has enriched us, we are proud to have been of his company and grateful for the privilege of his acquaintance.

The Masters' Lodge
Selwyn College
Cambridge — March 1985 — A. H. Cook

Author's Introductory Remarks

This book is the outcome of the attempt to approach the phenomenological problems of Geodesy, using, unlike many classical treatises based on fictitious reference systems, exclusively geometrical and physical objects and quantities embodied in the gravity field of the Earth itself, these being therefore accessible to actual observation. This approach leads directly to the three-dimensional domain which is the natural seat of the phenomena forming the object of the study of Geodesy.

The study was undertaken from the local point of view according to the *Nahewirkungsgesetz* (principle of action in the vicinity) of Hermann Weyl, using the method of vector and tensor analysis.

The collection of papers presented covers the period from 1950 to 1981 and has been ordered into six chapters.

The first chapter, entitled *Fundamentals of Intrinsic Geodesy* comprises a set of papers in which the foundations of Intrinsic Geodesy are given, making use of the natural observable coordinates latitude, longitude, and geopotential for which the fundamental metric tensor, the coefficients of connection, and the structure of the coordinate lines and surfaces are given.

The so-called first fundamental problem of Geodesy, of transfering the coordinates from one given point to another, is solved in three-dimensional space.

Application of the methods of intrinsic geodesy is also made to the study of the microgravitational field (the tidal field) of a satellite, or of a spacecraft in inertial motion, including the derivation of Ricci's coefficients of rotation which connect the eigenvectors of the tensor surfaces describing the field.

The second chapter, entitled *Structure of the Gravity Field and Laplace's Equation*, comprises two papers dealing with the curvature and torsion of the gravity field and a generalization of the famous Dalby's theorem which expresses, in an absolute form, Laplace's equation.

In the third chapter, entitled *Principles of Intrinsic Geodesy Applied to the Normal Reference Field*, the general equations established previously are applied to the case of the reference field endowed with rotational symmetry, e.g., Somigliana's ellipsoidal field. In this case the integrability conditions furnish the equations for the continuation of the field in space starting from the values assigned on the boundary surface.

The first fundamental problem of Geodesy for the transfer of the geographical coordinates and the potential along a given curve is solved. The fundamental parameters for the ellipsoidal field are computed.

In the fourth chapter, entitled *Mapping of the Actual Gravity Field onto the Normal Reference Field,* the correspondence between points of the surface of the Earth and the surface of an ellipsoid is generalized in three dimensions by establishing a one-to-one correspondence between points of the actual gravity field and points of the normal ellipsoidal reference field, assuming that the centre of mass of the Earth coincides with the centre of figure of the ellipsoidal field. Such correspondence is of Hirvonen's telluroidal type in which the potential is conserved. The aim of the mapping is the introduction of coordinates endowed with a simple metric tensor for easy computation; all the irregularities of the actual field are therefore removed to the modulus of deformation proper to the mapping.

A further aim of the mapping is to establish a rigorous definition of the anomalies based on analytical principles. The integrability conditions furnish, furthermore, a rigorous expression for the fundamental equation of Physical Geodesy and for the generalized Villarceau's equations relating the derivatives of the deflections of the vertical among themselves and the derivatives of the gravity anomaly. The conditions of harmonicity for the anomaly of the potential are also given.

A procedure for the adjustment of geodetic networks in the three-dimensional ellipsoidal model space is also given which generalises the method of variation of coordinates as used in the adjustment of bidimensional networks applying conformal representations.

One paper is devoted to the problem of conformal representations in the three-dimensional space. It is shown that it is impossible to introduce in conformal space a system of orthogonal coordinates having the transforms of the equipotential surfaces as one of the families of coordinate surfaces.

In the fifth chapter, entitled *Mapping Between Surfaces,* the mapping problem is approached from the local point of view by assuming that the quadratic form determining the modulus of deformation is assigned. The various types of representations are classified accordingly and the alterations induced in the curvatures are determined.

Some integral properties of the conformal representations relating the variation of the integral curvature with the flux of the gradient of the logarithm of the modulus of deformation are given.

In the sixth chapter, entitled *Propagation of a Light Path in Continuous Isotropic Refracting Media,* the geometric laws of propagation of a light ray in a continuous isotropic refracting medium are compared with the properties of conformal mapping in three dimensions.

The ideas expounded in the present collection of papers were first illustrated in the course of the General Assembly of the International Association of Geodesy in Oslo in 1948, where I had the good fortune to meet General Martin Hotine, who became greatly interested in the subject and developed independently a wealth of ideas that are condensed in his book *Mathematical Geodesy*, published in 1969.

The strong friendship that arose from the intimate collaboration with Martin Hotine was at the root of a number of Symposia originally held on Three-dimensional Geodesy and, then later, on Mathematical Geodesy. All Symposia were held, according to Martin's wish, in Italy (Venezia, 1959; Cortina d'Ampezzo, 1962; Torino, 1965; Trieste, 1969; Firenze, 1972; Siena, 1975; Assisi, 1978; Como, 1981). After Martin Hotine's death in 1968, the Symposia were dedicated to his memory.

The Symposia gave me the opportunity to become acquainted with a number of colleagues who greatly contributed to the development and extension of the original idea of Intrinsic Geodesy. I particularly wish to mention the most important contributions by Prof. Erik Grafarend of the University of Stuttgart and by Prof. Evangelos Livieratos of the University of Thessaloniki and their schools. I also wish to mention especially the fundamental considerations developed on the subject by Professor Nathaniel Grossman of the University of California, Los Angeles.

It was further my great good fortune to meet Dr. W. I. Reilly, who also became interested in the theory of Intrinsic Geodesy and has made important contributions to its application in practice. Being highly proficient in the Italian language, Dr. Reilly spontaneously undertook the task of translating my earliest papers, most of them written in my mother tongue, into English. I do not have sufficient words to thank him.

Special thanks are also due to Mrs. Maria Luisa Princivalli, my former student and now professor in the University of Trieste, who assisted me in carefully preparing the manuscripts and in the proofreading; and to Mrs. Ida C. Sbona, librarian of the Institute of Geodesy and Geophysics in the University of Trieste, who kindly offered to complete the bibliography from books in the Institute.

Antonio Marussi

Postscript. Antonio Marussi is no longer with us. His death has prevented him from giving the final touch to the manuscript, but his wife, Dolores Marussi, has spared no efforts to carry out his intentions faithfully and with great devotion, assisted by dedicated people, of whom Mrs. Ida C. Sbona has already been mentioned above. Particular thanks are due to Dipl.-Ing. Helmut Hornik, who has painstakingly checked and completed the bibliography, provided a list of abbreviations of institutions and publications quoted, and assisted in proofreading and in the final preparation of the manuscript for publication. The help and understanding of the Springer Publishing Company is gratefully acknowledged.

H. Moritz

Translator's Preface

The science of Geodesy has undergone far-reaching changes in the last half century. The impact of new technology, from electromagnetic distance measurements to the use of artificial satellites, has been great, and is still largely to be felt. These changes have forced the practitioners of the ancient art of Earth measurement to alter their way of thinking about the space that surrounds us, something fundamentally more difficult than absorbing a new technology.

A key influence in this modern change in geodetic thinking has been the work of Antonio Marussi, in his scientific publications from 1947 onwards, through his students and collaborators at the University of Trieste, and in the series of symposia on three-dimensional Geodesy which he organised with his great friend and collaborator, Martin Hotine. His influence on the latter, stemming from their first meeting at the General Assembly of the International Association of Geodesy in Oslo in 1948, was remarkable in itself, leading as it did to the ultimate publication of Hotine's *Mathematical Geodesy* in 1969.

Professor Marussi's strictly geodetic interests embrace two interrelated areas: (a) *Intrinsic Geodesy*, or the description of the Earth's gravity field in terms of coordinates natural to it – a modern extension and development of the differential geometry of Gauss, carried forward by a line of Italian mathematicians, amongst others; and (b) the theory of *two and three-dimensional representations* (or mapping projections), particularly as applied to the Earth's gravity field. Most of his papers from the fruitful decade of the 1950's were published in Italian, and some in journals of somewhat limited circulation, which tended to restrict their circulation among an international scientific community that was coming to rely on English as its principal language of communication.

The present volume results from a desire to present to the geodetic fraternity a significant cross-section of Professor Marussi's earlier works in English translation, as well as a selection of his later contributions published in English. The tone of the articles ranges from the concise style of the brief notes to the Accademia dei Lincei, on the one hand, to the frankly polemical of the opening speech to the first Symposium on Three-Dimensional Geodesy in Venice in 1959, on the other. This poses a small problem for the translator, for the Italian often carries rhetorical overtones in contrast to the more sober style that scientific English has inherited from the Royal Society of London. A more serious problem is the danger of imparting an awkwardness of diction, in no way justified by the original, by leaning towards a too-literal translation – a Scylla whose concomitant Charybdis is implicit in the Italian aphorism "Traduttore, traditore".

For today's reader, however, the problem is that he may suppose the author to have been tilting at windmills. One of the measures of the thoroughness of any revolution in ways of thinking about the universe is the difficulty which later generations have in imagining how it was before. It would not be amiss to cull the *Bulletin Géodésique* of the early 1950's to savour the nature of the debates recorded there, and the environment in which Professor Marussi's earlier works appeared.

Since first suggesting this project to Professor Marussi in 1980, I have to acknowledge assistance from a variety of sources: to the Director, Geophysics Division, Department of Scientific and Industrial Research, Wellington, New Zealand, for the provision of various services: to Miss Catherine Hourihan and Mrs. Winifred Esam for the typing of numerous manuscripts; to the Consiglio Nazionale delle Ricerche and the Istituto di Geodesia e Geofisica for hospitality at the University of Trieste in February–March 1982; and finally, the work of translation is offered as a tribute to Professor Marussi himself, in appreciation of his contribution to the science of Geodesy, and of a warm personal friendship that has grown and flourished over a period of 20 years and more, reinforced by those all-too-infrequent occasions when it has been possible to enjoy the congenial atmosphere of a Sunday at Sant'Elia in the company of Professor and Signora Marussi, an experience which those colleagues who have shared it will fully understand.

Wellington, August 1983 W. I. Reilly

Contents

I Fundamentals of Intrinsic Geodesy

1 From Classical Geodesy to Geodesy in Three Dimensions [1]

(Inaugural address at the Symposium on Three-Dimensional Geodesy, Venice, 16–18 July 1959).

There is no doubt that modern Geodesy, to be understood as the discipline that studies the earth from a geometric point of view, is the offspring of the experimental method. This it shares with many other disciplines, which, like ours, have their origin in the 17th century, and are nourished equally by experimental investigation and by mathematical speculation.

The experimental method is founded on the axiom that the only acceptable truths are those which experiment reveals or confirms, together with others which can be deduced from them by logical argument. Its way is by experiment, directed to the search for ever more recondite facts, and towards an increasing precision in observation.

However, the experimental and analytic study of natural phenomena would be a sterile exercise were it not followed by that synthesis which alone permits the inclusion of the immense variety of appearances in a vision of the whole in a few succinct statements of universal value, and which allows the prediction of facts that still belong to the future, and which elude our direct experimentation.

The 17th century not only fathered the experimental and analytical method, but also – and in this, too, following the teachings of Bacon – the inductive method. From the growing edifice of knowledge derived from observation, and from its critical evaluation, brilliant minds at an early stage distilled those universal truths which went to make up the marvellous structure of mechanics, and first among them the laws of statics and dynamics.

The principle of the impossibility of perpetual motion had already been divined by Leonardo a century before, but in the 17th century it found a more precise formulation in the work of Galileo and especially of Simon Stevin, the discoverer of the law of the parallelogram. Thence the principle of the conservation of energy took shape in the hands of René Descartes, Christian Huyghens and Galileo. Galileo stated the principle of inertia; Huyghens discovered the laws of centrifugal force, and Wallis, the admirable principle of virtual work.

The times thus ripened until the genius of Newton could bring together in one work the facts observed up to that time, and the principles which by then had

[1] Originally published as: Marussi A (1959) Dalla Geodesia classica alla Geodesia in tre dimensioni. Boll Geodes Sci Affini, anno XVIII: no 4

come to assume an ever more definite form. This work, moreover, by its elegance of thought and style, stands as an equal to the models of Greek scientific literature, and, by the power of its conception, impelled Lagrange to deem it "la plus haute production de l'esprit humain". Under Newton's hand the concepts of mass and force took solid shape, and the three fundamental principles of dynamics were added to the human heritage: the principle of inertia, the principle of proportionality between force and acceleration, and the principle of action and reaction.

Moreover, the force of gravity, that mysterious agent whose effects influence every act of our lives, but which still had no defined place either in geometric space or in human thought, not only found such a place, but straightway became that bridge to the final conquest of the physical world that surrounds us, uniting Physics with Astronomy, the earth with the heavens, and merging all harmoniously into the space that contains and regulates all the phenomena of the universe.

For millenia the sky had been for man a field for experimental research and observation. From the phenomena observed in it, his mind struggled in vain to discover those laws which were manifest in so many signs, but which with equal pertinacity escaped that complete synthesis which alone could quench his desire for knowledge. Astronomical cycles and empirical laws had long been known; but it was not possible to set them in order, one with the other; even less was it possible to link them with the facts of our earthly experience. The space which was their natural realm appeared as altogether different from that in which man dwells and acts.

In more recent times a great step had been accomplished: Copernicus had recognized the real nature of the solar system, and Kepler had discovered the precise kinematic laws of the movements of the heavenly bodies. Galileo, with his spyglass, had found around Jupiter a world like that foretold by Copernicus, and with this the geometry of planetary motion acquired shape in human consciousness.

But this was still a world of shadows which moved, bereft of body and of substance, across the screen of the heavens, and of the minds of these perspicacious astronomers – shadows subject certainly to laws, but to laws unconnected to each other, and not reducible to a single general principle.

At last, towards the end of the 17th century the times were ripe for the singling out of those laws that could lead the phenomena of physics and those of astronomy to the same universal principle, which would fill the void between earth and sky, and which would signal the final conquest of physical space, the sole home of all appearances.

But this step still needed the insight of a mind such as that of Isaac Newton. His insight was simple, but fearless and brilliant; it lay in extending the properties of the gravity field beyond the earth-space into the cosmic space with which it came to be united.

The world of man, flat as it had been up till then, became enriched in form and, at the same time, in content; an immense field opened up for research and experimentation.

The case of physical space is thus completely analogous to other cases which have frequently arisen in science, when the insight and fervid imagination of a brilliant mind can divine the existence of material entities, before they have been recognized in nature. We can recall how many chemical elements had been predicted by chemists, thanks to Mendeleev's laws, before they had in fact been discovered in nature; how the planet Neptune had first been discovered by the calculations of Leverrier, and only then identified in the sky by the telescope of Galle; how electromagnetic waves were first divined by Maxwell, and then found in nature by Hertz; and in the same way we can say that this has happened for space, of if we wish, for the third physical dimension.

After Newton, several centuries had to pass before man could put himself to the test in the third dimension, using machines of his own making, which lifted him above the flat earth on which he had lived for millennia. Today we stand as witnesses – I do not know whether fortunate or not – to this impressive moment that sees space furrowed by objects launched by our hands, and traversed by radiations produced and controlled to satisfy our insatiable desire to conquer and to know.

The discovery by Newton of physical space, and of gravity, the mysterious agent which occupies it, suddenly opened up a boundless field of research for the mathematicians: the first-born of this discovery was the theory of the potential, on which so much work of analysis has been done in the following centuries, along the road signposted with the names of Laplace, Green, and Stokes. It should no longer surprise us that here, also, abstract speculation had long preceded the results of experimentation, which still today is hard-pressed to keep pace. In effect, what we know today of the earth's gravity field owes very much more to work done at the desk, with pencil and paper, than to observations made with instruments in nature.

Let us then turn to our own discipline, to Geodesy. From the time that it was raised to the level of a discipline in the modern sense, it has been agitated by an age-old dilemma: constrained on the one side by theory, rich in productive and rigorous results, and on the other by experiment, rather less fruitful and not always of such excellence, it has pursued its path in a continuous search for an honourable compromise, thrust now this way and now that by the impulses which from time to time fashion imposes, but always hindered by fetters both material and psychological which have had a determining influence on its evolution.

The principal hindrance of a material nature has undoubtedly been the link which unites Geodesy to the practical necessities of life: the needs of map-making, surveying, navigation, and the military, all requiring above all else the representation of the Earth's surface on a plane. However, there is no possibility of escaping from this link, nor from the subjection which it entails, for it is for Geodesy a highly profitable service, and the only one which can supply it with the material means for progress; but as a counterweight it has imposed on it a one-sided and restricted view of its problems.

The other hindrance is rather of a psychological nature, and thus perhaps even more dangerous than the first. It is derived from the fact that for millennia man has considered the Earth a flat domain, or at least with two dimensions of overwhelming importance compared with the third. Only reason, but the reason of a mind such as that of Newton, has succeeded in taking account of the third dimension in all its consequences; but even this came late in the development of human knowledge and convictions. In fact, however, even after this conquest of incalculable scope, the mental habit of considering the earth as a flat surface, whether truly plane or curved, has clipped the wings of experimental Geodesy, constraining it to move in a world that is not its own, and thus to consider the third dimension as an undesirable intruder.

It happens even today that, while speculative Geodesy has for some time ranged in its natural environment of three-dimensional space, gathering a rich harvest, practical Geodesy lives in an unnatural and deficient world of only two dimensions. This forces it into the most absurd compromises, which, even if they have at times had great success in achieving immediate practical aims or results of first approximation, have yet impeded the path of steady progress.

The third dimension thus appears, in practical Geodesy, as an intruder in the flourishing paradise of the Geodesist, which comprises the surfaces of the ellipsoid and the geoid; and when this intruder comes to claim his proper rights, no efforts are spared to get rid of him as quickly as possible, and with the least trouble, by means of a weighty battery of corrections and reductions, which for two centuries has figured in every treatise on Geodesy.

Even today a good part of these treatises teach us from the first page that Geodesy is the discipline that studies the shape and dimensions of the Earth, when instead it would seem more reasonable to say that it is the discipline that studies the structure of the potential field of the gravity generated by the earth in space – a definition that should have *inter alia* the virtue of reconciling the present conflict between theoretical and operational Geodesy, eliminating the absurd distinction between dynamical Geodesy and geometrical Geodesy.

The dualism between operational Geodesy and theoretical Geodesy has generated the imaginary figure of the ellipsoid, which has appeared on the scene since the time of the celebrated controversy between the Newtonians and the Cassinians, endowed with the escutcheon of nobility in reward for having promoted the celebrated expeditions of the Academy of France to Peru and to Lapland.

From that time, this figure has dominated the scene in Geodesy with a double function and appearance. On the one hand it presents itself in the modest domestic role of the simplest of the mathematical surfaces which fits the figure of the Earth, and for which it is substituted whenever required or allowed by those practical needs to which our discipline is strongly subjected, as I have already remarked. We should recognize that the service rendered by the ellipsoid in this simple duty is worthy of our highest regard, if one only thinks that all geographic, topographic, and nautical maps and charts are based on it, and thus also are all the triangulations which cover great parts of the Earth.

Beyond this role, however, which we may consider a little ambiguous, but nevertheless very useful, the ellipsoid appears on the scene in Geodesy with the nobler role of a surface of comparison, when our research becomes more refined, and we can no longer be satisfied by the first approximation which it offers. Further, this model for comparison has been extended into space, thanks to the researches of Pizzetti, of Somigliana, and of Lambert, who have determined quite rigorously the gravity field of which the ellipsoid is an equipotential surface. If we were in a position to establish a one-to-one correspondence between the points of space and those of the model field of the ellipsoid, we would also be in a position to determine at every point the departures, or deviations, or anomalies of the real field with respect to the model field – something which for too many reasons we are still far from being able to achieve.

In the present state of affairs, each of these deviations, departures, or anomalies is subject to the arbitrary definition of such a correspondence, and here again the remedy has been to escape from the third dimension and to refer each point in real space to a point on the ellipsoid of comparison, with all the arbitrariness and violation of logic that is so well known to us.

These are the most apparent advantages and disadvantages of the ellipsoid. In its use, however, lies a danger which is concealed, and therefore, the more damaging, and for which it must be held to account: the way in which it has warped the conceptual development of operational Geodesy, which has certainly suffered as a result of the historical insistence on posing its problems within an "ellipsoidal" framework. It is enough to realize that even today that which is called the "first fundamental problem of Geodesy" is the transport of latitude, longitude, and azimuth along ellipsoidal geodesics, and that, today, geodetic networks of continental dimensions are based on this.

If today we raise a voice against the indiscriminate use of the ellipsoid, we shall perhaps be accused as suspected revolutionaries. This accusation would not be appropriate, for already before us, and not just a few years before – exactly 130 in fact – was one who saw clearly the blind alley into which Geodesy was heading by clinging too closely to the imaginary figure of the ellipsoid, and to the even more unreal figures traced on it by the geodesic lines. He who warned against this danger, and raised a clear voice of protest – but one which remained a voice crying in the wilderness – was J. Ivory. He figured in a celebrated controversy of the time with Tiarks, then head of the British Trigonometrical Survey, over Dalby's method which related the convergence of the normal sections at two points on an ellipsoid with the convergence of the meridians of the corresponding image on the sphere of directions, a method which can readily be generalized, as I have shown, to the case of a surface differing from that of the ellipsoid, for example that of the geoid.

The controversy was very much involved with the delicate question of the intrinsic significance of Laplace's equation. This can be stated in a way that does not depend on the ellipsoid, or on how the geoid is represented on it, when we consider that this equation allows us to obtain the astronomical differences of longitude, which are absolute quantities, from other quantities equally absolute, such as the astronomic azimuths.

Ivory's words on this subject are quite clear: "We have advanced no vague reasoning about the well known properties of spheroidical (ellipsoidical) triangles and geodetical lines in a case where, in fact, there is neither any such triangle nor any such line". And again, disputing against Tiarks: "But he has entirely mistaken the nature of the problem and the difficulties that must be overcome in solving it. There is no spheroidical triangle; there is no geodetical line. The data consists of the latitudes, nearly equal, and the azimuths, of two distant stations, yet sufficiently near for one to be seen from the other." (Ivory 1828, p. 432) There is thus no possible doubt as to Ivory's thinking; but many years were to pass before his words would be understood for their true and profound value.

Now the times seem to be ripe when the third dimension, which rounds out the space in which we are immersed, has at last ceased to be an abstraction and has become a field of experimentation, not only for the forces and means of nature, but also for those of man. Today the electromagnetic waves transmitted and controlled by us permit the measurement of distances along their geodesics, the optical trajectories. Today objects launched by the hand of man to the highest levels, if not yet into interplanetary space, can already be used as powerful probes to study its physical properties, or can be used as trigonometrical points to study its geometrical properties. Today the most refined methods of observation have allowed us to bring close our own satellite, the moon, and to make it a part of the terrestrial space, rather than of cosmic space. Today we are already talking of measuring directly the gravity at high levels and of knowing the higher harmonic components of the field itself by studying the perturbations of the orbits of artificial satellites.

If the seventeenth century was witness to the conceptual discovery of the physical space that surrounds us, it has fallen to our century to be present at its experimental conquest. It is the task of our discipline to prepare the methods which will allow us to coordinate the new directions of experimental research that technical developments have put at our disposal; and I think that this may be the reason that has brought us together in the idyllic surroundings of the Fondazione Cini on the island of San Giorgio, to discuss those methods which have been in preparation for these past few years.

For these methods also, theory has preceded the practical requirements and perhaps too the experimental possibilities, but the aim that has guided the present investigation has been that of satisfying an aesthetic need rather than a practical one. The desire for beauty, elegance, and simplicity, has pointed out new paths for research more often than might be imagined. The pace of research would be retarded, or even brought to a halt, were it not for the potent drive supplied by aesthetic aspirations, which is at times the sole justification for the effort exerted by the investigator, and which raises it to the level of an art.

In the case under consideration, the aesthetic aim to be achieved was to escape completely from the parasitic figure of the ellipsoid, which, summoned three centuries ago to the geodetic confraternity as an honoured and useful guest, gradually assumed the garb of an unwelcome visitor.

But to be freed of the ellipsoid, we needed to find another standpoint on which to base our measurements, and our further investigations. There are four ways that present themselves for this purpose, and all have been diligently pursued.

What is needed first of all is an unequivocal method of identifying the points of space, and for this it is necessary to set up a system of coordinates as a first step. To satisfy precise logical requirements and at the same time to serve practical needs, these coordinates should fulfill the following conditions: firstly, they should be intrinsic, which is to say that they should be able to be defined by means of real elements of the terrestrial gravity field, without recourse to extraneous or fictitious elements which have no physical reality; secondly, they should be physically observable and measurable, something which follows from their intrinsic nature.

These coordinates are, on each equipotential surface, the astronomic latitude and longitude: intrinsic quantities, defined by means of the vertical (which has a physical reality), and by means of the direction of the polar axis, and the disposition of the plane of the Greenwich meridian, all having physical reality. They are directly measureable, as we all well know from the personal experience of earlier days. The third coordinate, which distinguishes one level surface from another, is the potential, also intrinsically defined (except for an arbitrary constant), and directly observable, by combining the operations of levelling and gravimetry.

The coordinate surfaces thus defined constitute a triple family; the coordinate lines, their intersections, constitute in turn a triple congruence: of the parallels, of the meridians, and of the isozenithal lines (along the latter neither latitude nor longitude vary).

This is the fundamental framework which serves to sustain that which we can call Intrinsic Geodesy; a framework which allows us to achieve at every point a moving trihedron to which to refer, by means of their components, all the vector, or more generally, tensor, quantities which have to be considered.

This moving trihedron, comprising three base vectors, and well defined at every point, is obtained naturally by taking infinitesimal displacements along each of the coordinate lines, and forming the ratio of these displacements with respect to the infinitesimal increment in the corresponding parameter; at every point of space it thus has the same relationship with respect to the tensor quantities, as the unit of measurement has with respect to the scalar quantities. If one leaves aside the absolute orientation in space, it is completely defined once the moduli of its three base vectors are known, as well as the three angles formed between pairs. Six scalar quantities are involved in all, and these can be considered as the components of a symmetrical double tensor, the fundamental metric tensor.

The transition from one trihedron to the neighbouring one requires the knowledge of a certain number of coefficients, which are also to be taken as variables from point to point, and which will allow us to express the differential of a base vector as a linear function of the base vectors themselves, and of the components of the considered displacement, so that these coefficients must have three indices. As such, their number would be $3^3 = 27$, but this is reduced, to begin with, to 18, since the coefficients are symmetric with respect to two of the indices, and this number is reduced finally to 16 as a result of the particular properties assumed by the coordinates which we have defined.

These coefficients, which are known as connection coefficients in relation to the particular function which they fulfil, that of interconnecting the moving trihedra, can be expressed alternatively in terms of the derivatives of the components of the fundamental metric tensor; but they can also be obtained directly from geodetic observations.

To convince oneself of this in a quite elementary manner, it will suffice to consider the generalization of the first fundamental problem of Geodesy: that of the transport of coordinates (and of azimuth) along a geodesic. How is this problem generalized in operational Geodesy? Its enunciation remains unchanged in form; only now the geodesic will be a space geodesic, and thus a segment of a straight line. The coordinates will no longer comprise two only, but three: latitude, longitude, and potential (or dynamic height). To the transport of azimuth must naturally be added that of zenith distance.

From this it is easy to see that the transport of these elements implies the consideration of the coefficients of connection; hence it is easy to infer that measurement of length, azimuth, and zenith distance for any space geodesic connecting points of known latitude, longitude, and dynamic height, will allow us to set up equations from which to derive the desired connection coefficients.

We have seen how the introduction of the three geoidic coordinates will allow us to define at every point in space a triad of base vectors which are not, however, in general either unit or mutually orthogonal. It could be asked whether it would not be easier to realize such a triad directly, taking as the first vector that defined by the vertical, as the second that directed to the North, and as the third that directed toward the East. This triad, which we shall call astronomical, and which in turn supplies us with a reference trihedron at every point of space, satisfies, as well as the first one, the axiomatic conditions we have laid down: intrinsic definition, and accessibility to observation. Why, then, not use this triad, instead of the geoidic, which is not orthogonal, and whose definition is not as immediate?

In effect, there is no reason not to do so, and it has been done brilliantly by Brigadier Hotine. It is only that in such a case it is inevitably necessary to give up the use of general coordinates, in as much as the astronomical triad is not the principal triad of any triple system of surfaces in space. In other words, there exists in space no triple system of surfaces which intersect each other at every point along the congruences of the north and east lines, and the line of force. To be persuaded of this, it is sufficient to recall the enunciation of Dupin's celebrated theorem: if these surfaces were to exist, they would need to intersect along the respective lines of curvature; hence the north and east directions would need to be principal directions at every point of the field, which is obviously absurd if the field is not endowed with axial symmetry. This therefore excludes the use of such triads in combination with coordinates related to them. However, it in no way excludes the use to which they can profitably be put, as a base for Cartesian or polar coordinates related to one triad at a time which can be taken as fundamental. It is these latter coordinates which naturally present themselves when, as is becoming usual today, we measure geodesic distances in space using new electronic or optical apparatus, and determine the astronomic azimuth and the reciprocal zenith distances at both ends of the geodesic.

One can also naturally raise here the problem of connecting an astronomical triad with another of the same kind close to it. Here too occur particular connection coefficients, in this case known as Ricci's rotation coefficients, which, when multiplied by the base vectors and by the Cartesian components of the displacement, yield the differentials of the base vectors of the astronomical triad itself. Such coefficients are all expressed simply by the curvature parameters of the gravity field.

A third procedure, conceived by Brigadier Hotine, sets out from the following fundamental observation. By developing the whole theory of the Intrinsic

Geodesy of the gravity field by the tensor method, one obtains formulas which are valid in any coordinate system. Why not then choose from these coordinates a convenient system, which is both apt for practical needs, and which provides coordinates of a type and of values not too different from the accustomed ones? Thus Brigadier Hotine suggests that we take up coordinates based upon an ellipsoid; according to this, each point of space is defined rather simply by the latitude and the longitude of the normal projected from the point onto the ellipsoid, as well as by its distance from it.

These are the coordinates that have been used by Brigadier Hotine and by his worthy collaborator Dr. Brazier, in what we can justly call the first practical application of these new methods in Geodesy, which deals specifically with the adjustment of an extensive network of measurements of varying types made in the White Sands Area of Colorado. In this regard we should be grateful to Mr. Whitten of the U.S. Coast & Geodetic Survey who has not only made available this valuable set of measurements, but has also arranged that the adjustment calculations should be done on the electronic calculating machines of the same Coast & Geodetic Survey.

The methods which I have briefly outlined are connected naturally with others which have been worked out by other investigators, by Prof. Väisälä, by Dr. O'Keefe, by Dr. Sodano, and which take account of the geodetic use of objects at a great height above the Earth's surface, such as aircraft in flight at a high level, illuminated rockets launched from aircraft or from missiles, or artificial satellites; or which are based on the method proposed by Banakiewicz for the geodetic use of the total eclipses of the sun.

All these methods, which envisage the use of points very distant from the Earth's surface, and thus observation along space geodesics, exclude the use of the classical procedures of Geodesy, and impose different requirements, in a three-dimensional space, for treating the problems which now arise.

The fourth and last way is that which adopts the oldest coordinate system of all, the geocentric Cartesian system, with the origin at the centre of gravity of the Earth. It is a system familiar to astronomers, which has been used continuously for several centuries in this field, and which lies, for example, at the base of all theories of the tides and of the motion of the Moon. It was of course not unknown to geodesists, to whom it was of use especially in theoretical investigations. But it failed to meet one of the fundamental requirements we have laid down, that of accessibility to measurement. The centre of gravity of the Earth has been up till now a pure abstraction, and proof against either direct or indirect observation.

Today matters are no longer so. Artificial satellites describe orbits which we can observe with our instruments, and these orbits are geometrically connected with the Earth's centre of gravity; and so this ceases to be a pure abstraction, and enters as a physical reality into operational Geodesy.

This fourth system of spatial coordinates, which Dr. Markowitz has pointed out to us, along with its method of use, today satisfies the premises which we have laid down, and thus gains full rights of citizenship in operational Geodesy, as it already has in Astronomy.

The most modern views have thus taken us back along the oldest roads. The oldest coordinates, the astronomical latitude and longitude, used by the Greek and Arab astronomers, supplanted for several centuries by the ellipsoidal, acquire new rights. The plane triangles, formed by space geodesics, which our forefathers observed with the sextant, and which the theodolite and ellipsoidal Geodesy had relegated to the historical museum for the last century and a half, reappear under another form, but in substance unchanged, in the most modern technique of trilateration; and so too reappear geocentric coordinates.

The cycle closes on itself and is renewed, and reveals to our eyes new and distant horizons, which it will be our task to explore.

If our symposium brings a contribution in this direction, we shall have reason to be satisfied.

Bibliography

Ivory J (1828) On the method in the trigonometrical survey for finding the difference of longitude of two stations very little different in latitude. Philos Mag, Ann Philos, December 1828:432

2 Foundations of Intrinsic Geodesy[1]

Introduction

The intrinsic geometry of a surface is concerned with its study by reference only to elements measurable by a hypothetical being constrained to stay upon it. These elements are obviously the linear lengths of the arcs of any curves drawn upon the surface, and also the tangential curvatures of these, as well as the angle between two distinct arcs. Specifically excluded, however, are all the linear and angular measurements which entail leaving the surface, and as a result, there is no possibility of determining the Cartesian coordinates in space of a point on the surface, nor the direction of the normal to the surface.

For such a study, we usually choose Gauss' procedure, introducing on the surface a double family of coordinate lines; and such a study coincides, as we know, with that of a Riemannian manifold of two dimensions.

The practical application of these concepts in Geodesy is found in the operations of base-line measurement and of triangulation (excluding trigonometric levelling), for which it is not necessary to make measurements that leave the level surface on which we imagine that we work.

The intrinsic study of a surface can be expanded, by supposing that we know how to make angular measurements in the surrounding space, and in particular to determine the direction of the normal to the surface with respect to a system of versors (which by convention we shall call absolute), but without permitting linear measurements outside the surface. This new concept also finds practical application in Geodesy, and particularly in Geodetic Astronomy, where the direction of the vertical is referred to an absolute system of versors related to the celestial vault.

To define the direction of the considered vertical we make use as usual of the representative sphere of directions, on which we consider an ordinary system of geographic coordinates, taking the Earth's axis as the polar axis. The intrinsic study of the surface can then be done by defining on it a system of geographic coordinates, making each point on it correspond to the spherical image of the surface normal.

Having thus introduced geographic coordinates, we shall be able to use them in the same manner as ordinary curvilinear coordinates. If the first fundamental form of the surface is given in these coordinates, then the second fundamental

[1] Originally published as: Marussi A (1951) Fondamenti di Geodesia intrinseca. Pubbl Comm Geodet Ital Terza Ser Mem 7: 47pp

form of the surface is completely determined, thus assuring the significance of these coordinates in the third dimension.

The interest of such investigations for Geodesy is evident, inasmuch as they involve only quantities accessible to direct measurement, and do not require the intervention of any arbitrary simplifying hypothesis. Several geodesists have occupied themselves with working out the geodetic applications, and among them Reina and Viterbi; but of the greatest importance are the investigations of Pizzetti and Mineo, of which the essential features are summarized here. To Mineo in particular we owe a synthesis in which these principles are founded upon an organic and clear basis.

These investigations have, however, remained limited in geodetic applications to one particular surface, such as, for example, the equipotential surface of gravity which coincides with mean sea level (where it exists), and is called the geoid. Moreover, we note that this surface belongs to the family of level surfaces of the terrestrial potential, a family which enjoys the property that if any one of them be known, external to all attracting mass, then all other external surfaces are known. This leads to the extension of the theory to the whole of space, by means of the introduction of a third coordinate, equally intrinsic, in addition to latitude and longitude.

To prepare for such an extension, it is useful to begin with the absolute study of an isolated surface, or of the entire family to which it belongs, without introducing any coordinates. For this the vectorial and homographic methods of Burali-Forti, Burgatti, Boggio, Marcolongo, etc. are excellently suited. The theory of surfaces following such methods is to be found extensively treated in various treatises; but the fine investigations of Prof. Caldonazzo which are largely reported here, as well as those of the writer, have shown how these absolute methods can be of unparalleled clarity and utility, also in the study of a potential field in both its geometric and mechanical aspects. The investigations of Caldonazzo and Manacorda have shown how to make use of homographic methods also in the determination of discontinuities in the potential derivatives, which arise when crossing surfaces of discontinuity in the density of the attracting material. This problem is also of considerable importance in Geodesy, in the study of the internal gravity field of mountains and of the Earth. In such absolute investigations, the consideration of two particular homographies is of fundamental importance: that of Eötvös, comprising the spatial derivatives of the gravity vector, and the generalized homography of Burali-Forti, comprising the spatial derivatives of the versor normal to the equipotential surface. The simple connection between these two homographies characterizes the relationships between the mechanical and geometric properties of the second order of the field, providing a picture of this that is extremely clear and synthetic.

By projecting vectors and homographies onto a local system of versors, of which one is directed along the vertical, the second normal to this towards north, the third normal to both and towards east (the local astronomical system), we can then make a study of the field by appropriate measurements of scalar quantities.

The various systems of local astronomical versors which could be considered at each point of space are, however, unconnected with each other. From the properties of an entity referred to one of these, one could not easily deduce the properties referred to a neighbouring system, and moreover, it will be shown that the local astronomical systems are not apt for the definition of a triple system of coordinate surfaces in space.

When we wish to study the general properties of the field from other than a purely local viewpoint, we need to define a general system of coordinates for all the surrounding space, so as to have intrinsic character, and for practical reasons to be accessible to direct measurement. It has already been seen that on an isolated surface these coordinates are the geographic, and it is natural to take as the third intrinsic coordinate that which distinguishes one equipotential surface from another, i.e. the potential difference, when we take one of the surfaces (the geoid) as a reference surface. The potential difference between one surface and another measures the so-called dynamic height difference, obtained from levelling, with the introduction of an inessential factor ($1/g_{0.45}$) in which $g_{0.45}$ is the normal value of gravity at 45° of latitude at sea level ($g_{0.45} = 980.6294$ Gal from the Tables of Cassinis calculated on the basis of Somigliana's theory).

We have thus defined in space a triple family of coordinate surfaces, of which one is composed of the equipotential surfaces, another of the surfaces of equal latitude, and the third of those of equal longitude.

These surfaces intersect along a triad of congruences of curves, the coordinate curves, which are known as the geodetic meridians and parallels, and the isozenithal lines.

For the geodetic parallels and meridians the fundamental theorem of Pizzetti holds, by which they are related to the north and east directions by means of the surface homography of Burali-Forti. At the same time, the direction of the tangent to the isozenithal line is the null direction of the generalized homography of Burali-Forti. The study of these problems, limited to a selected point of space, constitutes the local geodetic approach.

The system of general coordinates, introduced in this way, now permits the transport of any geometric or mechanical property of the field from one point to another in space. This can be done analytically by means of the vector calculus, or better by means of the absolute differential calculus of Ricci and Levi-Civita, also known as the tensor calculus. We have with the first a greater intuition in the symbolism, and in the second, the advantage of operating directly with scalar quantities accessible to direct measurement, or readily deducible from such. This is the general geodetic approach, which allows us to confront the problems of *Intrinsic Geodesy* which require, as is most usual, consideration of geometric and mechanical entities at distant points of space.

In this chapter the problem outlined has been attacked with complete generality, following the directions which have been discussed. First comes the study of the field by the absolute approach, from which follows that by the local approach, using Cartesian and geographical coordinates. Having thus established the fundamental principles, we go on to develop the general approach using intrinsic coordinates, and both vector and tensor methods, which has the advantages (at least in the writer's view) of combining the intuitive character of the formulae with the ease of translating these into terms of scalar quantities accessible to direct measurement[2]. The treatment concludes with a short chapter – in no way intended to be exhaustive – on some immediate applications.

We should note also that the investigations discussed here can easily be applied to the gravity field of Somigliana, for which one equipotential surface is an ellipsoid. If we take this to be the International Ellipsoid, and the equatorial gravity as that determined by Silva, we have the "normal field" defined for practical geodetic application by Cassinis, and accepted internationally as such at the Stockholm meeting of the International Association of Geodesy and Geophysics in 1930. We thus have available a *comparison field* with which the actual gravity field of the Earth can be compared; but such a field need not at all be involved in conceptual or numerical developments into which only the actual results of measurements should be introduced.

Finally we should note how Geodesy in very recent times has come to extend its geometric measurements into space. While in the past all linear measurements were carried out along arcs of surface geodesics, or were reduced to these, we tend today to measure distances along lines of space by optical or radioelectrical means. Also from this point of view, a broadening of geodetic theory would seem to be of some interest.

A Absolute Approach

1. The Vectors $\boldsymbol{g}$ *and* $\boldsymbol{n}$; *the Homographies* $\mathscr{W}$ *and* σ; *the Gravimetric Gradient* $\mathbf{G}$; *the Isogravitational Surfaces*

Avoiding any reference to coordinates, we shall be able to define at any point P of space, external to the attracting masses, the value W of the potential of gravity – Newtonian potential plus potential of the centrifugal force – the versor $\boldsymbol{n}$ normal to the equipotential surface Σ passing through it, taken positively towards the exterior (towards W decreasing), and the vector $\boldsymbol{g}$ which represents in magnitude, direction, and sense the gravity

$$\boldsymbol{g} = \operatorname{grad} W = -g\boldsymbol{n} . \tag{1.1}$$

Alongside the vectors $\boldsymbol{g}$ and $\boldsymbol{n}$, characteristic respectively of the mechanical and geometric nature of the field, we shall consider the homographies

$$\mathscr{W} = \frac{dg}{dP} = \frac{d}{dP} \operatorname{grad}_P W \tag{1.2}$$

$$\sigma = \frac{dn}{dP} \tag{1.3}$$

which represent their spatial derivatives. For reasons that will be made clear in what follows, these homographies will be called the *homography of Eötvös,* and

[2] Such a method has been adopted, in a slightly different form, in the splendid treatise on "Tensor Calculus and Applications" (Finzi and Pastori 1949)

the *generalized homography of Burali-Forti,* respectively. It is evident that the homography $\mathscr{W}$ of Eötvös can synthesize all the mechanical characteristics of the second order of the field, and that of Burali-Forti the geometric properties. The analytic relationship which we are about to establish between these two homographies synthesizes in turn the link between its mechanical and geometric structure.

We have immediately from the equation defining $\boldsymbol{g}$

$$\boldsymbol{n} = -\frac{\boldsymbol{g}}{g} \tag{1.4}$$

$$\mathscr{W} = -g\sigma - H(\operatorname{grad} g, \boldsymbol{n}) \tag{1.5}$$

$$\sigma = -\frac{1}{g}\{\mathscr{W} + H(\operatorname{grad} g, \boldsymbol{n})\} \tag{1.6}$$

where H is the symbol of the dyadic of Gibbs ($H(\boldsymbol{a},\boldsymbol{b})\boldsymbol{x} = \boldsymbol{a} \times \boldsymbol{x} \cdot \boldsymbol{b}$). In these two fundamental relationships appears the new vector

$$\boldsymbol{G} = \operatorname{grad} g\,, \tag{1.7}$$

which takes the name of *gravimetric gradient*; it is determined easily by noting that

$$\operatorname{grad}\boldsymbol{g} \times \boldsymbol{g} = 2g \operatorname{grad} g = 2K\frac{d\boldsymbol{g}}{dP}\boldsymbol{g} = -2gK\mathscr{W}\boldsymbol{n} = -2g\mathscr{W}\boldsymbol{n}\,, \tag{1.8}$$

since $\mathscr{W}$, as the spatial derivative of a gradient, is a dilatation, equal to its own conjugate $K\mathscr{W}$

$$\mathscr{W} = K\mathscr{W} \tag{1.9}$$

and hence

$$\boldsymbol{G} = \operatorname{grad} g = -\mathscr{W}\boldsymbol{n} \tag{1.10}$$

and (1.6) can now also be written as

$$\sigma = -\frac{1}{g}\mathscr{W} + \frac{1}{g}H(\mathscr{W}\boldsymbol{n},\boldsymbol{n})\,, \tag{1.11}$$

while (1.5) shows that the knowledge of the homography σ is not sufficient to determine $\mathscr{W}$, but that it is also necessary to know the scalar g and the vector $\boldsymbol{G}$.

We have then the following important relationships, which follow directly from the properties of the operator K, and of the products of homographies:

$$\mathscr{W} = K\mathscr{W} = -gK\sigma - H(\boldsymbol{n},\boldsymbol{G}) \tag{1.12}$$

$$K\sigma = -\frac{1}{g}[\mathscr{W} + H(\boldsymbol{n},\boldsymbol{G})]\,, \tag{1.13}$$

from which immediately

$$K\sigma\boldsymbol{n} = 0 \tag{1.14}$$

and hence

$$\mathscr{W}\sigma = (K\,\mathscr{W})\,\sigma = -g\sigma^2 - H(K\sigma\boldsymbol{G},\boldsymbol{n}) = -g(K\sigma)\,\sigma\,, \tag{1.15}$$

which will be of use in what follows.

The expression (1.13) found for $K\sigma$ allows us to decompose the homography σ into two dyads; if we take $\boldsymbol{a}$ and $\boldsymbol{b}$ as two orthogonal surface versors whatever on Σ, we have

$$\sigma = -\frac{1}{g}[H(\mathscr{W}\boldsymbol{a},\boldsymbol{a}) + H(\mathscr{W}\boldsymbol{b},\boldsymbol{b})]\,. \tag{1.16}$$

The surfaces $g=$ const, whose differential equation is given obviously by

$$\boldsymbol{G}\times dP = 0 \tag{1.17}$$

take the name of *isogravitational surfaces*.

2. The Homography σ_Σ of Burali-Forti: Curvatures

Since $\boldsymbol{n}$ is a unit vector, $\boldsymbol{n}\times\boldsymbol{n}=1$, $\boldsymbol{n}\times\sigma=0$; and the homography σ transforms any vector whatever into a vector tangent to the equipotential surface Σ considered.

Limiting for the moment the study to tangential versors, it is convenient to use instead of the spatial homography σ, the surface homography σ_Σ defined by

$$\begin{aligned}\sigma_\Sigma &= \sigma - H(\boldsymbol{n},\sigma\boldsymbol{n}) = \sigma[1 - H(\boldsymbol{n},\boldsymbol{n})]\\ &= -\frac{1}{g}\{\mathscr{W} - H(\mathscr{W}\boldsymbol{n},\boldsymbol{n}) + gH(\boldsymbol{n},\sigma\boldsymbol{n})\}\,.\end{aligned} \tag{2.1}$$

The surface homography σ_Σ is a dilatation; in fact, from the lamellarity of the field, by

$$\operatorname{rot}\boldsymbol{g} = \operatorname{rot}\operatorname{grad}W = -g\operatorname{rot}\boldsymbol{n} + \boldsymbol{n}\wedge\boldsymbol{G} = 0 \tag{2.2}$$

and by

$$\operatorname{rot}_\Sigma\boldsymbol{n} = \operatorname{rot}\boldsymbol{n} - \boldsymbol{n}\wedge\sigma\boldsymbol{n} \tag{2.3}$$

it follows that

$$\boldsymbol{n}\wedge\boldsymbol{G} = g\boldsymbol{n}\wedge\sigma\boldsymbol{n} \tag{2.4}$$

$$\operatorname{rot}_\Sigma\boldsymbol{n} = 0\,. \tag{2.5}$$

The homography σ_Σ, of which $\operatorname{rot}_\Sigma\boldsymbol{n}$ represents the double of the vector, is hence a dilatation, in order for such a vector to be zero. Considered in space, however, the homography σ is not a dilatation. It is thus a degenerate homography of the first kind, since it transforms any space vector into a vector parallel to the plane tangent to the surface $W=$ const at the point considered, as we have already seen; and thus we have $I_3\sigma=0$.

We shall now say that two versors $\boldsymbol{x}$ and $\boldsymbol{y}$ are conjugate with respect to the homography σ_Σ if

$$\sigma_\Sigma \boldsymbol{x} \times \boldsymbol{y} = \sigma_\Sigma \boldsymbol{y} \times \boldsymbol{x} = 0\,; \tag{2.6}$$

the commutation of $\boldsymbol{x}$ and $\boldsymbol{y}$ is possible, σ_Σ being a dilatation, in virtue of Jacobi's theorem. The autoconjugate versors of the homography represent the asymptotic directions of the surface; while its principal directions $\boldsymbol{c}_1$ and $\boldsymbol{c}_2$ are given obviously by the equation

$$\sigma \boldsymbol{c} \times \boldsymbol{n} \wedge \boldsymbol{c} = 0\,; \tag{2.7}$$

these yield the curvature versors (versors of the lines of curvature of Σ), and their investigation coincides with the investigation of the two united directions of σ:

$$\sigma \boldsymbol{c}_1 = \frac{1}{r_1}\boldsymbol{c}_1\,, \qquad \sigma \boldsymbol{c}_2 = \frac{1}{r_2}\boldsymbol{c}_2 \tag{2.8}$$

in which r_1 and r_2 represent the principal radii of curvature. It follows immediately that:

$$I_1\sigma = \frac{1}{r_1} + \frac{1}{r_2} = \mathrm{H} \qquad \text{(mean curvature of } \Sigma) \tag{2.9}$$

$$I_2\sigma = \frac{1}{r_1 r_2} = \mathrm{K} \qquad \text{(total or Gaussian curvature)} \tag{2.10}$$

$$I_1^2 - 2I_2 = \frac{1}{r_1^2} + \frac{1}{r_2^2} = \mathrm{C} \quad \text{(curvature of Casorati)}\,. \tag{2.11}$$

We have also

$$\operatorname{div} \boldsymbol{n} = I_1 \sigma_\Sigma = \mathrm{H}\,. \tag{2.12}$$

3. *The Frenet Equations*

Denoting by $\boldsymbol{t} = \dfrac{dP}{ds}$ the tangent versor to a curve of arc s traced on Σ, and putting

$$\boldsymbol{b} = \boldsymbol{n} \wedge \boldsymbol{t} \tag{3.1}$$

such that $(\boldsymbol{n}, \boldsymbol{t}, \boldsymbol{b})$ is a right-handed unit triad; denoting further by $\boldsymbol{N}$ and $\boldsymbol{B}$ respectively the versors of the principal normal and of the binormal to the curve, oriented so that the trihedron $(\boldsymbol{t}, \boldsymbol{N}, \boldsymbol{B})$ is in turn right-handed, we have the equations of Frenet (or of Serret)

$$\frac{d\boldsymbol{t}}{ds} = \frac{\boldsymbol{N}}{\rho}\,; \qquad \frac{d\boldsymbol{N}}{ds} = -\frac{\boldsymbol{t}}{\rho} - \frac{\boldsymbol{B}}{\tau}\,; \qquad \frac{d\boldsymbol{B}}{ds} = \frac{\boldsymbol{N}}{\tau}\,, \tag{3.2}$$

in which $\dfrac{1}{\rho}$ and $\dfrac{1}{\tau}$ represent respectively the first curvature or flexion (absolute) and the second curvature or torsion (absolute) of the given curve. Hence we have the generalized equations of Frenet

$$\left.\begin{aligned}\frac{dn}{ds} &= \sigma(t) = -\frac{t}{\rho_n}+\frac{b}{\tau_g}\\ \frac{dt}{ds} &= \frac{N}{\rho} = \frac{b}{\rho_g}+\frac{n}{\rho_n}\\ \frac{db}{ds} &= -\frac{t}{\rho_g}-\frac{n}{\tau_g}\end{aligned}\right\} \tag{3.3}$$

having put

$$\left.\begin{aligned}\frac{1}{\rho_n} &= \frac{\cos\theta}{\rho} && \text{(normal curvature)}\\ \frac{1}{\rho_g} &= \frac{\sin\theta}{\rho} && \text{(tangential or geodetic curvature)}\\ \frac{1}{\tau_g} &= \frac{d\theta}{\tau ds} && \text{(geodetic torsion)}\end{aligned}\right\} \tag{3.4}$$

and where θ (the angle between N and n) is defined by the following relationships

$$\left.\begin{aligned}N &= n\cos\theta+b\sin\theta\\ B &= n\sin\theta-b\cos\theta\end{aligned}\right\}. \tag{3.5}$$

Projecting the first equation of (3.3) onto t and b we have the following expressions for the normal curvature and for the geodetic torsion

$$\frac{1}{\rho_n} = -\sigma(t)\times t\,; \qquad \frac{1}{\tau_g} = \sigma(t)\times b\,. \tag{3.6}$$

The kinematic vector of Darboux for the surface triad (n,t,b) is then given by

$$\omega_\Sigma = \frac{n}{\rho_g}-\frac{t}{\tau_g}-\frac{b}{\rho_n}. \tag{3.7}$$

If the given curve is a geodesic, $\frac{1}{\rho_g} = 0$, and the kinematic vector becomes tangential; on the other hand it lies in the plane normal to the curve, if the latter is a line of curvature $\left(\frac{1}{\tau_g} = 0\right)$; and it is directed along b if the curve is simultaneously a line of curvature and a geodesic.

4. *The Lines of Force*

By now considering the spatial homography σ, the vector

$$\sigma n = -\frac{1}{g}(\mathscr{W}n+G\times n\cdot n) = \frac{1}{g}\operatorname{grad}_\Sigma g \tag{4.1}$$

normal to n, obviously has as modulus the curvature $\frac{1}{\mathrm{R}} = \mathrm{F}$ of the line of force; and the plane $(n, \sigma n)$ is the osculating plane to the line itself. From Eqs. (2.2) and (2.4) we then have

$$\mathrm{rot}\,\boldsymbol{n} = \boldsymbol{n} \wedge \sigma \boldsymbol{n} = \frac{\boldsymbol{B}'}{\mathrm{R}} \tag{4.2}$$

if by $\boldsymbol{B}'$ we denote the versor of the binormal to the line of force. The vector $\mathrm{grad}_\Sigma\, g$ is the surface component of grad g.

Denoting analogously by N' the versor of the principal normal to the line of force, we then have

$$\sigma \boldsymbol{n} = \frac{\boldsymbol{N}'}{\mathrm{R}} = \frac{1}{g}\mathrm{grad}_\Sigma\, g = (\mathrm{rot}\,\boldsymbol{n}) \wedge \boldsymbol{n}\ , \tag{4.3}$$

from which

$$\mathrm{F} = \frac{1}{\mathrm{R}} = \frac{1}{g}\mathrm{grad}_\Sigma\, g \times \boldsymbol{N}' = \frac{1}{g}\boldsymbol{G} \times \boldsymbol{N}'\ . \tag{4.4}$$

5. Bruns' Equation; Expression for $\Delta_2 g$

We now take account of the mechanical nature of the vector $\boldsymbol{g}$, given by the two relationships

$$\mathrm{rot}\,\boldsymbol{g} = 0\ ; \qquad \mathrm{div}\,\boldsymbol{g} = 2\,\omega^2 - 4\,\pi f \delta \tag{5.1}$$

in which ω is the angular velocity of rotation of the Earth [$\omega = 72.9213 \times 10^{-6}$ (cgs)], f the universal constant of attraction [$f = 6.664 \times 10^{-8}$ (cgs)] and δ the spatial density of the attracting matter (in the case of external surfaces $\delta = 0$). It has already been seen that from $\mathrm{rot}\,\boldsymbol{g} = 0$ it follows that

$$\mathrm{rot}\,\boldsymbol{n} = \frac{1}{g}\boldsymbol{n} \wedge \boldsymbol{G}\ . \tag{5.2}$$

Analogously we have, from the expression of $\mathrm{div}\,\boldsymbol{g} = \mathrm{div}\,\mathrm{grad}\, W = \Delta_2 W$

$$\mathrm{div}\,\boldsymbol{n} = \frac{4\,\pi f \delta - 2\,\omega^2}{g} - \frac{1}{g}\boldsymbol{G} \times \boldsymbol{n}\ . \tag{5.3}$$

Recalling that $\mathrm{div}\,\boldsymbol{n} = I_1 \sigma = \mathrm{H}$ we have also

$$\mathrm{grad}\, g \times \boldsymbol{n} = 4\,\pi f \delta - 2\,\omega^2 - g\mathrm{H}\ ; \tag{5.4}$$

and under this form the relationship is known as Bruns' equation; it is of fundamental importance in the so-called free air reduction of gravity measurements.

Passing to the study of the second-order operators, we note first of all the very remarkable equation

$$v\,\frac{d\,\mathscr{W}}{dP} = \mathrm{grad}\,\mathscr{W} = \Delta'\,\mathrm{grad}\, W = -4\,\pi f\,\mathrm{grad}\,\delta\ ; \tag{5.5}$$

and furthermore we have, after a somewhat complex deduction

$$\Delta_2 g = \mathrm{div}\,\boldsymbol{G} = g(\mathrm{C} + \mathrm{F}^2) + 4\,\pi f\,\mathrm{grad}\,\delta \times \boldsymbol{n}\ , \tag{5.6}$$

a rather important equation which links the Laplacian of the modulus of gravity with the gravity itself, the curvature C of Casorati of the equipotential surface, the first curvature $F = 1/R$ of the line of force, and the normal derivative of density. Such an expression for $\Delta_2 g$ has been given by Caldonazzo in the equivalent form

$$\Delta_2 g = g I_1(K\sigma \cdot \sigma) - \operatorname{grad} \Delta_2 W \times \boldsymbol{n} \,. \tag{5.7}$$

6. *Invariants of $\mathscr{W}$ and of σ; Null Directions of σ and of $K\sigma$*

Passing now to the calculation of the invariants of the homographies σ and $\mathscr{W}$ we have first of all, as has already been stated,

$$\left.\begin{array}{ll} I_1\sigma = I_1\sigma_\Sigma = \mathrm{H}\,; & I_2\sigma = I_2\sigma_\Sigma = \mathrm{K}\,; \\ I_3\sigma = 0\,; & (I_1\sigma)^2 - 2I_2\sigma = \mathrm{C} \end{array}\right\} . \tag{6.1}$$

For the first invariant of $\mathscr{W}$ we can restate Bruns' equation in the form

$$I_1\mathscr{W} = \Delta_2 W = 2\omega^2 - 4\pi f\delta = -g\mathrm{H} - \boldsymbol{G}\times\boldsymbol{n} \tag{6.2}$$

and then

$$I_2\mathscr{W} = g^2(\mathrm{K}-\mathrm{F}^2) + g\mathrm{H}\cdot\boldsymbol{G}\times\boldsymbol{n} = -g^2(\mathrm{F}^2+\mathrm{C}+\mathrm{K}) - g\mathrm{H}\Delta_2 W\,. \tag{6.3}$$

So too, recalling that

$$I_3\mathscr{W} = (\mathscr{W}\boldsymbol{c}_1)\wedge\mathscr{W}\boldsymbol{c}_2\times\mathscr{W}\boldsymbol{n}\,, \tag{6.4}$$

we obtain

$$I_3\mathscr{W} = g^2\left(\frac{\boldsymbol{N}'\times\boldsymbol{c}_1}{r_2\mathrm{R}}\boldsymbol{c}_1 + \frac{\boldsymbol{N}'\times\boldsymbol{c}_2}{r_1\mathrm{R}}\boldsymbol{c}_3 - \frac{\boldsymbol{n}}{r_1 r_2}\right)\times\boldsymbol{G}\,. \tag{6.5}$$

By an easy calculation it is seen that we have for the homography $R\sigma$

$$R\sigma = H([K\sigma\boldsymbol{c}_1]\wedge[K\sigma\boldsymbol{c}_2],\boldsymbol{n})\,, \tag{6.6}$$

which allows us to assert that the zero direction of σ is given by $(K\sigma\boldsymbol{c}_1)\wedge(K\sigma\boldsymbol{c}_2)$. As we shall see, this is the direction of the isozenithal line. It has already been seen that, on the other hand, $\boldsymbol{n}$ is the null direction of $K\sigma$.

7. *Discontinuities of $\mathscr{W}$, σ, and* Grad g

As is known, the potential of gravity and its gradient are continuous on crossing a surface S of density discontinuity, but the homography $\mathscr{W}$ is not. In particular, denoting by $\boldsymbol{n}'$ the outward normal to the surface S which encloses, in its interior, masses of density δ, and denoting by $\mathscr{W}_e$ the homography $\mathscr{W}$ external to S, and by $\mathscr{W}_i$ that internal,

$$\mathscr{W}_i - \mathscr{W}_e = [\mathscr{W}] = -4\pi f\delta H(\boldsymbol{n}',\boldsymbol{n}')\,. \tag{7.1}$$

In particular, applying the homography $[\mathscr{W}]$ to $\boldsymbol{n}$ we have

$$[\operatorname{grad} g] = (\operatorname{grad} g)_i - (\operatorname{grad} g)_e = 4\pi f\delta\boldsymbol{n}'\times\boldsymbol{n}\cdot\boldsymbol{n}'\,. \tag{7.2}$$

This relationship allows us to find the jumps undergone by the gravimetric elements on crossing a surface, and the jumps in the curvature of the lines of force and their principal normals.

To find the jumps undergone by all the other gravimetric elements, it is sufficient to determine the discontinuities of the homography σ

$$\sigma_i - \sigma_e = [\sigma] = \frac{4\pi f \delta}{g} H(\boldsymbol{n}', \boldsymbol{n}'_\Sigma) \tag{7.3}$$

having denoted by $\boldsymbol{n}'_\Sigma$ the component along Σ of the versor $\boldsymbol{n}'$.

Nothing further will be said concerning discontinuities; but those who wish to study this question more deeply should turn to the important studies of Caldonazzo, who has examined the discontinuities of the third derivatives of the volumetric potential across an equipotential surface which separates elements of differing density, and of Manacorda, who has equally studied the behaviour, up to the third derivatives, of the potential of a single layer crossing the generating layer.

B Local Astronomical Approach

8. The Local System and the Absolute System of Astronomical Versors

Having thus examined in detail the structure of a gravity field from an absolute viewpoint, in both its mechanical and geometric aspects, it will now be convenient, in the study of the Earth's gravity field, to refer the absolute entities considered to a Cartesian system of axes, linked to a point in space as origin, and to the vertical and the Earth's axis for orientation. The projections of the various entities considered onto this system of Cartesian axes will yield scalars which in many cases can be measured directly by geodetic operations carried out at the point P in space (or in its immediate vicinity). A typical instrument which can be used for this purpose is the torsion balance (of different kinds), as will be seen in the applications.

The Cartesian system, which we shall call the local astronomical, is defined thus: the axis of x^3 (orthometric height), represented by the versor $\boldsymbol{i}_3$, is directed along the normal $\boldsymbol{n}$ to the equipotential surface through P, in an outward sense; the axis of x^1, represented by the versor $\boldsymbol{i}_1$, is perpendicular to the previous one, and is contained in the plane of the astronomical meridian (the plane $[\boldsymbol{i}_3, \boldsymbol{a}_3]$), if $\boldsymbol{a}_3$ is the versor which represents the Earth's rotation, directed positively towards the north; the axis of x^2, represented by the versor $\boldsymbol{i}_2$, is perpendicular to both the others, and directed positively toward the east. The triad $(\boldsymbol{i}_1, \boldsymbol{i}_2, \boldsymbol{i}_3)$ so defined is right-handed, in tribute to the fact that in Geodesy azimuths are measured in a clockwise sense.

Introducing a left-handed triad $(\boldsymbol{a}_1, \boldsymbol{a}_2, \boldsymbol{a}_3)$ of orthogonal versors, which we shall conventionally call absolute in the sense that, in contrast to those previously defined, it will hold for any point of the field, and such that $\boldsymbol{a}_3$ is directed along the Earth's axis positively to the north, $\boldsymbol{a}_1$ is contained in the plane of the astronomical meridian at Greenwich, and $\boldsymbol{a}_2$ perpendicular to the other two, we shall then have

$$\left.\begin{aligned} &\boldsymbol{a}_1 = \boldsymbol{a}_3 \wedge \boldsymbol{a}_2; \qquad \boldsymbol{a}_3 = \boldsymbol{a}_1 \wedge \boldsymbol{a}_3; \qquad \boldsymbol{a}_3 = \boldsymbol{a}_2 \wedge \boldsymbol{a}_1 \\ &\boldsymbol{i}_1 = -\sin\varphi\cos\lambda \cdot \boldsymbol{a}_1 - \sin\varphi\sin\lambda \cdot \boldsymbol{a}_2 + \cos\varphi \cdot \boldsymbol{a}_3 \\ &\boldsymbol{i}_2 = \quad -\sin\lambda \cdot \boldsymbol{a}_1 \qquad + \cos\lambda \cdot \boldsymbol{a}_2 \\ &\boldsymbol{i}_3 = \cos\varphi\cos\lambda \cdot \boldsymbol{a}_1 + \cos\varphi\sin\lambda \cdot \boldsymbol{a}_2 + \sin\varphi \cdot \boldsymbol{a}_3 \end{aligned}\right\} \tag{8.1}$$

$$\left.\begin{aligned} \boldsymbol{a}_1 &= -\sin\varphi\cos\lambda\cdot\boldsymbol{i}_1-\sin\lambda\cdot\boldsymbol{i}_2+\cos\varphi\cos\lambda\cdot\boldsymbol{i}_3 \\ \boldsymbol{a}_2 &= -\sin\varphi\sin\lambda\ \ \boldsymbol{i}_1+\cos\lambda\cdot\boldsymbol{i}_2+\cos\varphi\sin\lambda\cdot\boldsymbol{i}_3 \\ \boldsymbol{a}_3 &= \cos\varphi\ \ \boldsymbol{i}_1 \qquad\qquad\qquad +\sin\varphi\cdot\boldsymbol{i}_3 \end{aligned}\right\}. \tag{8.2}$$

The absolute triad is chosen as left-handed, in tribute to the convention of measuring longitudes in an anticlockwise sense.

By differentiating we arrive at the following relationships:

$$\left.\begin{aligned} d\boldsymbol{i}_1 &= \qquad\qquad -\boldsymbol{i}_2\sin\varphi\,d\lambda-\boldsymbol{i}_3\,d\varphi \\ d\boldsymbol{i}_2 &= \boldsymbol{i}_1\sin\varphi\cdot d\lambda \qquad\qquad -\boldsymbol{i}_3\cos\varphi\,d\lambda \\ d\boldsymbol{i}_3 &= \boldsymbol{i}_1\,d\varphi \qquad +\boldsymbol{i}_2\cos\varphi\,d\lambda \end{aligned}\right\} \tag{8.3}$$

and further, for the kinematic vector

$$\omega = -\boldsymbol{i}_1\cos\varphi\,d\lambda+\boldsymbol{i}_2\,d\varphi-\boldsymbol{i}_3\sin\varphi\cdot d\lambda\,. \tag{8.4}$$

9. Cartesian Components

For simplicity in writing we shall now denote the partial derivatives of the potential W with respect to x^1, x^2, x^3 by the following symbols:

$$\frac{\partial W}{\partial x^r} = W_r\,; \qquad \frac{\partial^2 W}{\partial x^r\,\partial x^s} = W_{rs} \tag{9.1}$$

and we shall also put

$$w_{rs} = -\frac{1}{g}\,W_{rs}\,. \tag{9.2}$$

With these symbols, we have immediately

$$\boldsymbol{i}_3 = \boldsymbol{n} = -\frac{1}{g}\operatorname{grad} W \tag{9.3}$$

$$g = -W_3 \tag{9.4}$$

$$\boldsymbol{g} = \operatorname{grad} W = -g\boldsymbol{i}_3 = W_3\boldsymbol{i}_3 \tag{9.5}$$

and hence

$$\left.\begin{aligned} \mathscr{W}\boldsymbol{i}_1 &= W_{11}\boldsymbol{i}_1+W_{12}\boldsymbol{i}_2+W_{13}\boldsymbol{i}_3 \\ \mathscr{W}\boldsymbol{i}_2 &= W_{21}\boldsymbol{i}_1+W_{22}\boldsymbol{i}_2+W_{23}\boldsymbol{i}_3 \\ \mathscr{W}\boldsymbol{i}_3 &= W_{31}\boldsymbol{i}_1+W_{32}\boldsymbol{i}_2+W_{33}\boldsymbol{i}_3 \end{aligned}\right\} \tag{9.6}$$

$$W_{rs} = \mathscr{W}\boldsymbol{i}_r\times\boldsymbol{i}_s = \mathscr{W}\boldsymbol{i}_s\times\boldsymbol{i}_r\,. \tag{9.7}$$

Using matrix symbolism, we shall thus also write

$$\mathscr{W} = \left\|\begin{matrix} W_{11} & W_{12} & W_{13} \\ W_{21} & W_{22} & W_{23} \\ W_{31} & W_{32} & W_{33} \end{matrix}\right\| = \|\,W_{rs}\,\| \tag{9.8}$$

which justifies the name of homography of Eötvös proposed for $\mathscr{W}$, inasmuch as it was this geodesist who gave the first systematic consideration to the applications of the second derivatives of the potential to geodetic and geophysical problems.

From the mechanical definition of W we have

$$I_1 \mathscr{W} = \operatorname{div}\boldsymbol{g} = W_{11} + W_{22} + W_{33} = 2\omega^2 - 4\pi f \delta . \tag{9.9}$$

We have also directly

$$\operatorname{grad} g = \boldsymbol{G} = -\mathscr{W}\boldsymbol{i}_3 = -W_{31}\boldsymbol{i}_1 - W_{32}\boldsymbol{i}_2 - W_{33}\boldsymbol{i}_3 . \tag{9.10}$$

Considering the dyad $\frac{1}{g} H(\mathscr{W}\boldsymbol{i}_3, \boldsymbol{i}_3)$ we then have

$$\frac{1}{g} H(\mathscr{W}\boldsymbol{i}_3, \boldsymbol{i}_3) \equiv \begin{Vmatrix} 0 & 0 & -w_{31} \\ 0 & 0 & -w_{32} \\ 0 & 0 & -w_{33} \end{Vmatrix} . \tag{9.11}$$

From the relationship which yields σ

$$\sigma = \frac{d\boldsymbol{i}_3}{dP} = -\frac{1}{g}\{\mathscr{W} - H(\mathscr{W}\boldsymbol{i}_3, \boldsymbol{i}_3)\} , \tag{9.12}$$

$$\sigma \equiv \begin{Vmatrix} w_{11} & w_{12} & 0 \\ w_{21} & w_{22} & 0 \\ w_{31} & w_{32} & 0 \end{Vmatrix} . \tag{9.13}$$

We have also

$$\sigma_\Sigma = \begin{Vmatrix} w_{11} & w_{12} \\ w_{21} & w_{22} \end{Vmatrix} . \tag{9.14}$$

From these relationships we immediately recover the values of the invariants

$$\left.\begin{aligned} I_1\sigma &= \mathrm{H} = w_{11} + w_{22} \\ I_2\sigma &= \mathrm{K} = w_{11} w_{22} - w_{12}^2 \end{aligned}\right\} . \tag{9.15}$$

10. Differential Relationships for the Astronomical Versors

From the definition of the local astronomical triad, we have directly the following expressions for the spatial derivatives of the base astronomical versors

$$\frac{d\boldsymbol{i}_1}{dP} = \begin{Vmatrix} 0 & -w_{12}\operatorname{tg}\varphi & -w_{11} \\ 0 & -w_{22}\operatorname{tg}\varphi & -w_{12} \\ 0 & -w_{32}\operatorname{tg}\varphi & -w_{13} \end{Vmatrix} ; \qquad \frac{d\boldsymbol{i}_2}{dP} = \begin{Vmatrix} w_{12}\operatorname{tg}\varphi & 0 & -w_{12} \\ w_{22}\operatorname{tg}\varphi & 0 & -w_{22} \\ w_{32}\operatorname{tg}\varphi & 0 & -w_{32} \end{Vmatrix} \tag{10.1}$$

$$\sigma = \frac{d\boldsymbol{i}_3}{dP} = \begin{Vmatrix} w_{11} & w_{12} & 0 \\ w_{21} & w_{22} & 0 \\ w_{31} & w_{32} & 0 \end{Vmatrix} \tag{10.2}$$

and hence

$$\left.\begin{aligned} \operatorname{div} \boldsymbol{i}_1 &= -w_{22} \operatorname{tg} \varphi - w_{13} \\ \operatorname{div} \boldsymbol{i}_2 &= \quad w_{12} \operatorname{tg} \varphi - w_{23} \\ \operatorname{div} \boldsymbol{i}_3 &= \quad w_{11} + w_{22} \end{aligned}\right\} . \tag{10.3}$$

It is now known that we can always put

$$\frac{d\boldsymbol{i}_r}{dP} = -\boldsymbol{i}_r \wedge \gamma \tag{10.4}$$

where γ is the rotational homography of Pensa.

With an easy calculation we have

$$\gamma = \left\| \begin{matrix} -w_{12} & w_{11} & -w_{12} \operatorname{tg} \varphi \\ -w_{22} & w_{12} & -w_{22} \operatorname{tg} \varphi \\ -w_{32} & w_{13} & -w_{32} \operatorname{tg} \varphi \end{matrix} \right\| . \tag{10.5}$$

We have further

$$\left.\begin{aligned} \operatorname{rot} \boldsymbol{i}_1 &= (-w_{12} + w_{23} \operatorname{tg} \varphi) \boldsymbol{i}_1 + w_{11} \boldsymbol{i}_2 - w_{12} \operatorname{tg} \varphi \cdot \boldsymbol{i}_3 \\ \operatorname{rot} \boldsymbol{i}_2 &= -w_{22} \boldsymbol{i}_1 + (w_{12} + w_{23} \operatorname{tg} \varphi) \boldsymbol{i}_2 - w_{22} \operatorname{tg} \varphi \cdot \boldsymbol{i}_3 \\ \operatorname{rot} \boldsymbol{i}_3 &= -w_{23} \boldsymbol{i}_1 + w_{13} \boldsymbol{i}_3 \end{aligned}\right\} \tag{10.6}$$

$$I_1 \gamma = -w_{23} \operatorname{tg} \varphi . \tag{10.7}$$

11. The Local Astronomical Triads are Not Principal Triads of Any Triply Orthogonal System

The necessary and sufficient conditions that the local astronomical triads should be principal triads of a triply orthogonal system of surfaces, are given by

$$I_1 \gamma = 0 ; \qquad \operatorname{rot} \boldsymbol{i}_t \times \boldsymbol{i}_t = 0 ; \qquad \gamma \boldsymbol{i}_t \times \boldsymbol{i}_t = 0 . \tag{11.1}$$

All these conditions lead to the two very simple relations:

$$w_{12} = 0 ; \qquad w_{23} = 0 \tag{11.2}$$

which characterize a field endowed with rotational symmetry about the Earth's axis. Since in general such relations are not satisfied, we can assert that the congruences of the north and east lines, and the lines of force, do not in general belong to a triply orthogonal system; and they are not therefore suitable for defining a general system of coordinates in space.

12. United Directions; Curvatures

The united directions of σ (or, what is the same, of σ_Σ) are given by the equation

$$\sigma \boldsymbol{c} = k \boldsymbol{c} \quad \text{and thus} \quad (\sigma - k) \boldsymbol{c} = 0 \tag{12.1}$$

or again

$$(\sigma c) \wedge c = 0 . \tag{12.2}$$

Putting

$$c = \cos\theta \cdot i_1 + \sin\theta \cdot i_2 , \tag{12.3}$$

where θ is the astronomic azimuth of c, the last vector equation yields directly

$$w_{12} \mathrm{tg}^2 \theta + (w_{11} - w_{22}) \mathrm{tg}\, \theta - w_{12} = 0 \tag{12.4}$$

$$\mathrm{tg}\, 2\theta_{1,2} = 2 \frac{w_{12}}{w_{11} - w_{22}} , \tag{12.5}$$

while the relation

$$I_2(\sigma_\Sigma - k) \equiv \left\| \begin{matrix} w_{11} - k & w_{12} \\ w_{21} & w_{22} - k \end{matrix} \right\| = 0 \tag{12.6}$$

tells us that k is the solution of the quadratic equation

$$k^2 - k(w_{11} + w_{22}) + (w_{11} w_{22} - w_{12}^2) = 0 \tag{12.7}$$

i.e.

$$k^2 - \mathrm{H} k + \mathrm{K} = 0 . \tag{12.8}$$

As is well known, the versors c_1 and c_2, to which the values θ_1 and θ_2 of the astronomical azimuths correspond, are the curvature versors of the surface, while the values k_1 and k_2 which are solutions of Eq. (12.7) provide the principal radii of curvature.

If a is any tangential versor

$$a = \cos\alpha \cdot i_1 + \sin\alpha \cdot i_2 , \tag{12.9}$$

$$\left(\frac{1}{\rho_n}\right)_\alpha = \sigma a \times a = w_{11} \cos^2\alpha + 2 w_{12} \sin\alpha \cos\alpha + w_{22} \sin^2\alpha \tag{12.10}$$

yields the curvature of the geodesic of azimuth α. Its torsion is given by

$$\left(\frac{1}{\tau_g}\right)_\alpha = \sigma a \times n \wedge a = \frac{1}{2}(w_{22} - w_{11}) \sin 2\alpha + w_{13} \cos 2\alpha . \tag{12.11}$$

In particular, for geodesics of astronomic azimuth zero or a right angle (tangent to the North and East lines of the surface) we have

$$\begin{aligned} &\left(\frac{1}{\rho_n}\right)_{\text{North}} = w_{11} ; \qquad \left(\frac{1}{\tau_g}\right)_{\text{North}} = w_{12} \\ &\left(\frac{1}{\rho_n}\right)_{\text{East}} = w_{22} ; \qquad \left(\frac{1}{\tau_g}\right)_{\text{East}} = - w_{12} . \end{aligned} \tag{12.12}$$

From

$$\sigma n = \frac{N'}{\mathrm{R}} = \mathrm{F}N' , \tag{13.1}$$

where N' is the principal normal of the line of force, we have

$$N' = \cos \eta \boldsymbol{i}_1 + \sin \eta \boldsymbol{i}_2 = \boldsymbol{R}(w_{13}\boldsymbol{i}_1 + w_{23}\boldsymbol{i}_2) \tag{13.2}$$

$$\mathrm{tg}\, \eta = \frac{w_{23}}{w_{13}} ; \qquad \mathrm{F}^2 = w_{13}^2 + w_{23}^2 , \tag{13.3}$$

where η is then the azimuth of the osculating plane.

Analogously to the line of force, characterized by the equation $\boldsymbol{t} = \dfrac{dP}{ds} = \boldsymbol{i}_3$ we can consider the curves characterized by the relationships

$$\boldsymbol{t} = \boldsymbol{i}_1 ; \qquad \boldsymbol{t} = \boldsymbol{i}_2 ; \tag{13.4}$$

these curves take the names of North and East lines respectively.

Considering first the North line, and in Eq. (3.3) introducing ds_1 as an infinitesimal displacement along $\boldsymbol{i}_1$

$$\left.\begin{aligned}
&\left(\frac{1}{\rho_n}\right)_{\mathrm{North}} = \frac{d\varphi}{ds_1} = w_{11} ; \qquad \left(\frac{1}{\tau_g}\right)_{\mathrm{North}} = \cos\varphi \frac{d\lambda}{ds_1} = w_{12} \\
&\left(\frac{1}{\rho_g}\right)_{\mathrm{North}} = -\sin\varphi \frac{d\lambda}{ds_1} = -w_{12}\mathrm{tg}\,\varphi = -\left(\frac{1}{\tau_g}\right)_{\mathrm{North}} \mathrm{tg}\,\varphi \\
&\frac{1}{\rho_{\mathrm{North}}} = \sqrt{w_{11}^2 + w_{12}^2 \mathrm{tg}^2 \varphi} \\
&\frac{a\lambda}{ds_1} = \sqrt{\left(\frac{1}{\tau_g}\right)_{\mathrm{North}}^2 + \left(\frac{1}{\rho_g}\right)_{\mathrm{North}}^2} = \frac{w_{12}}{\cos\varphi}
\end{aligned}\right\} . \tag{13.5}$$

These equations yield, as well as the normal curvature along the astronomic meridian $\left(\dfrac{1}{\rho_n}\right)_{\mathrm{North}}$ and the geodetic torsion along the same $\left(\dfrac{1}{\tau_g}\right)_{\mathrm{North}}$, also the geodetic curvature of the North lines $\left(\dfrac{1}{\rho_g}\right)_{\mathrm{North}}$, and their absolute curvature $\dfrac{1}{\rho_{\mathrm{North}}}$, expressed by means of the second derivatives of the potential.

Analogous equations can naturally be found for the East lines, and we should have

$$\left.\begin{aligned}
&\left(\frac{1}{\rho_n}\right)_{\mathrm{East}} = \cos\varphi \frac{d\lambda}{ds_2} = w_{22} ; \qquad \left(\frac{1}{\tau_g}\right)_{\mathrm{East}} = -\frac{d\varphi}{ds_2} = -w_{12} \\
&\left(\frac{1}{\rho_g}\right)_{\mathrm{East}} = -\sin\varphi \frac{d\lambda}{ds_2} = -w_{22}\mathrm{tg}\,\varphi = -\left(\frac{1}{\rho_n}\right)_{\mathrm{East}} \mathrm{tg}\,\varphi \\
&\frac{1}{\rho_{\mathrm{East}}} = \frac{d\lambda}{ds_2} = \frac{w_{22}}{\cos\varphi}
\end{aligned}\right\} . \tag{13.6}$$

It should be noted that the East lines are the plane sections of the surfaces Σ with planes perpendicular to the Earth's axis of rotation; since in fact $d\boldsymbol{i}_2 \times \boldsymbol{a}_3 = 0$, their absolute torsion is zero, and they have, from Bonnet's theorem, $\theta = \varphi$.

By considering any curve whatever, from a comparison of Eq. (3.3) with Eq. (8.3), and putting A as the azimuth of the curve, and $\boldsymbol{t} = \boldsymbol{i}_1 \cos A + \boldsymbol{i}_2 \sin A$, we have the fundamental property

$$\frac{1}{\rho_g} = -\sin\varphi \frac{d\lambda}{ds} + \frac{dA}{ds}. \tag{13.7}$$

From this the equation of a geodesic takes one of two equivalent forms

$$\frac{1}{\rho_g} = 0; \qquad dA = \sin\varphi \cdot d\lambda \tag{13.8}$$

the latter being analogous to the Laplace equation.

The following equation should also be noted

$$\frac{d\varphi}{ds}\cos A + \frac{d\lambda}{ds}\cos\varphi \sin A = \frac{1}{\rho_n}. \tag{13.9}$$

14. Discontinuities

Concerning the discontinuities of the second derivatives of the potential (the potential itself and its gradient are continuous everywhere), it is only necessary to project on the axes $\boldsymbol{i}_1, \boldsymbol{i}_2, \boldsymbol{i}_3$ the relationships [Eq. (7.1)], putting for $\boldsymbol{n}'$, the external normal to the surface of discontinuity

$$\boldsymbol{n}' = \sin\nu\cos\alpha \cdot \boldsymbol{i}_1 + \sin\nu\sin\alpha \cdot \boldsymbol{i}_2 + \cos\nu \cdot \boldsymbol{i}_3 \tag{14.1}$$

with ν the zenith distance, and α the astronomic azimuth. We then have

$$\left.\begin{aligned}
[W_{11}] &= -4\pi f\delta \sin^2\nu\cos^2\alpha \\
[W_{22}] &= -4\pi f\delta \sin^2\nu\sin^2\alpha \\
[W_{33}] &= -4\pi f\delta \cos^2\nu \\
[W_{12}] &= -4\pi f\delta \sin^2\nu\sin\alpha\cos\alpha \\
[W_{13}] &= -4\pi f\delta \sin\nu\cos\nu\cos\alpha \\
[W_{23}] &= -4\pi f\delta \sin\nu\cos\nu\sin\alpha
\end{aligned}\right\} \tag{14.2}$$

C Local and General Geodetic Approach

15. Geodetic Meridians and Parallels: Pizzetti's Theorem

The homography σ_Σ causes to correspond to every surface versor $\boldsymbol{a}$, another surface vector $\boldsymbol{a}'$; the versor $\boldsymbol{b}$ normal to $\boldsymbol{a}'$ will thus be called the conjugate of $\boldsymbol{a}$ with respect to σ_Σ, and since the homography σ_Σ is a dilatation, we shall have from Jacobi's commutation theorem

$$a' \times b = \sigma a \times b = \sigma b \times a = b' \times a = 0\,, \tag{15.1}$$

from which it is seen that the conjugation relationship is reciprocal.

Let us now consider the versor conjugate to the versor $\boldsymbol{i}_1$, which characterizes on Σ the astronomical north direction, and let us denote it by $\boldsymbol{i}'_2$; we have

$$\sigma \boldsymbol{i}_1 \times \boldsymbol{i}'_2 = \sigma \boldsymbol{i}'_2 \times \boldsymbol{i}_1 = 0\,. \tag{15.2}$$

Analogously, considering the versor conjugate to $\boldsymbol{i}_2$, which we shall call $\boldsymbol{i}'_1$, we shall have

$$\sigma \boldsymbol{i}_2 \times \boldsymbol{i}'_1 = \sigma \boldsymbol{i}'_1 \times \boldsymbol{i}_2 = 0\,. \tag{15.3}$$

Working through the calculations, and by convention taking always the positive value of the root, we have, setting

$$\left.\begin{aligned} \boldsymbol{i}'_1 &= \cos\mu \cdot \boldsymbol{i}_1 + \sin\mu \cdot \boldsymbol{i}_2 \\ \boldsymbol{i}'_2 &= \cos\pi \cdot \boldsymbol{i}_1 + \sin\pi \cdot \boldsymbol{i}_2 \end{aligned}\right\} \tag{15.4}$$

$$\sigma \boldsymbol{i}'_1 = \frac{I_2 \sigma}{\sqrt{w_{12}^2 + w_{22}^2}} \boldsymbol{i}_1\,; \qquad \sigma \boldsymbol{i}'_2 = \frac{I_2 \sigma}{\sqrt{w_{11}^2 + w_{12}^2}} \boldsymbol{i}_2 \tag{15.5}$$

$$\operatorname{tg}\mu = -\frac{w_{12}}{w_{22}}\,; \qquad \operatorname{tg}\pi = -\frac{w_{11}}{w_{12}}\,. \tag{15.6}$$

Since $\sigma dP = d\boldsymbol{n}$, the equations written above show that, for a displacement of P along $\boldsymbol{i}'_1$, the variation of $\boldsymbol{n}$ is entirely parallel to $\boldsymbol{i}_1$; and that, for a displacement of P along $\boldsymbol{i}'_2$, the variation of $\boldsymbol{n}$ is entirely parallel to $\boldsymbol{i}_2$.

For these reasons, $\boldsymbol{i}'_1$ and $\boldsymbol{i}'_2$ will be said to define the directions of the geodetic meridian and parallel on Σ; and we shall be able to state *Pizzetti's Theorem*: the tangents to the geodetic meridian (parallel) are conjugate, with respect to the homography σ_Σ, to the East (North) cardinal lines.

The equations written above thus yield the azimuths μ and π of the geodetic meridian and parallel; their angle γ is given by

$$\operatorname{tg}\gamma = -\frac{1}{w_{12}} \frac{\mathrm{K}}{\mathrm{H}}\,. \tag{15.7}$$

The normal curvatures and the geodetic torsions of the geodetic meridian and parallel are given straight away by

$$\left(\frac{1}{\rho_n}\right)_{\text{merid.}} = \frac{-w_{22}\mathrm{K}}{w_{12}^2 + w_{22}^2}\,; \qquad \left(\frac{1}{\tau_g}\right)_{\text{merid.}} = \frac{w_{12}\mathrm{K}}{w_{12}^2 + w_{22}^2} \tag{15.8}$$

$$\left(\frac{1}{\rho_n}\right)_{\text{parall.}} = \frac{-w_{11}\mathrm{K}}{w_{11}^2 + w_{12}^2}\,; \qquad \left(\frac{1}{\tau_g}\right)_{\text{parall.}} = \frac{-w_{12}\mathrm{K}}{w_{11}^2 + w_{12}^2}\,. \tag{15.9}$$

16. Geographic Coordinates; the Three Fundamental Forms

We are now in a position to introduce the geographic coordinates in an organic form, as intrinsic coordinates of the surface. Their intrinsic definition is given by the fundamental relationship

$$dn = i_1 d\varphi + i_2 \cos\varphi\, d\lambda \tag{16.1}$$

which will be called the *third fundamental form* of Σ. Since, as we have seen, only φ various along i_1', and only λ along i_2', we shall be able to write quite generally

$$dP = i_1' \sqrt{E}\, d\varphi + i_2' \sqrt{G}\, d\lambda \tag{16.2}$$

which we shall call the *first fundamental form* of Σ. $\sqrt{E}$ and $\sqrt{G}$ are essentially positive coefficients, which yield the metric of Σ in the chosen system of geographic coordinates.

Giving explicit form to the relationship

$$dn = \sigma dP \tag{16.3}$$

we immediately obtain

$$\left.\begin{aligned} E &= \frac{w_{12}^2 + w_{22}^2}{K^2}\,; \qquad G = \frac{w_{11}^2 + w_{12}^2}{K^2} \cos^2\varphi \\ F &= \frac{H}{K^2}\, w_{12} \cos\varphi \end{aligned}\right\} \tag{16.4}$$

in which we set

$$F = i_1' \times i_2' \sqrt{EG} = \sqrt{EG} \cos\gamma \tag{16.5}$$

with

$$ds^2 = dP^2 = E d\varphi^2 + 2F d\varphi d\lambda + G d\lambda^2 . \tag{16.6}$$

Returning to the first fundamental form of Σ

$$dP = i_1' \sqrt{E}\, d\varphi + i_2' \sqrt{G}\, d\lambda \tag{16.7}$$

and expressing it by the astronomical versors instead of the geodetic ones, we have

$$dP = (\sqrt{E} \cos\mu\, d\varphi + \sqrt{G} \cos\pi\, d\lambda) i_1 + (\sqrt{E} \sin\mu\, d\varphi + \sqrt{G} \sin\pi\, d\lambda) i_2 . \tag{16.8}$$

If we now observe that

$$\sqrt{G} \cos\pi = \sqrt{E} \sin\mu \cos\varphi \tag{16.9}$$

and if we put

$$\left.\begin{aligned} &\sqrt{E} \cos\mu = -D\,; \qquad \sqrt{G} \sin\pi \cos\varphi - D'' \\ &\sqrt{G} \cos\pi = \sqrt{E} \sin\mu \cos\varphi = -D' \end{aligned}\right\}, \tag{16.10}$$

the foregoing is written

$$-dP = (D d\varphi + D' d\lambda) i_1 + \frac{D' d\varphi + D'' d\lambda}{\cos\varphi} i_2 , \tag{16.11}$$

and under this aspect we shall call it the *second fundamental form of the surface*.

In Cartesian coordinates linked to the axes i_1 and i_2 we shall have

$$dP = i_1 dx^1 + i_2 dx^2 \tag{16.12}$$

and hence

$$\left.\begin{aligned} dx^1 &= -D\,d\varphi - D'\,d\lambda \\ dx^2 &= \frac{-D\,d\varphi - D''\,d\lambda}{\cos\varphi} \end{aligned}\right\} \tag{16.13}$$

and furthermore

$$\frac{\partial P}{\partial\varphi} = i_1'\sqrt{E} = -D\boldsymbol{i}_1 - \frac{D'\boldsymbol{i}_2}{\cos\varphi}; \qquad \frac{\partial P}{d\lambda} = i_2'\sqrt{G} = -D'\boldsymbol{i}_1 - \frac{D''\boldsymbol{i}_2}{\cos\varphi}. \tag{16.14}$$

Differentiating the relationship

$$dP \times \boldsymbol{n} = 0 \tag{16.15}$$

we then have

$$-dP \times d\boldsymbol{n} = \boldsymbol{n} \times d^2P \tag{16.16}$$

and thus

$$-dP \times d\boldsymbol{n} = \boldsymbol{n} \times d^2P = D\,d\varphi^2 + 2D'\,d\varphi\,d\lambda + D''\,d\lambda^2. \tag{16.17}$$

We note that the differential equations

$$dx^1 = 0; \qquad dx^2 = 0 \tag{16.18}$$

are characteristic of the East and North lines respectively.

17. Relationships Between the First and Second Forms

As is seen, the coefficients of the second fundamental form are connected to those of the first in finite terms; this occurs in the particular case being studied, in as much as the chosen intrinsic geographic coordinates have a significance in the space in which the surface Σ is immersed. In the generality of cases, however, the first fundamental form is characteristic of the intrinsic geometry of the surface; the second is characteristic of the relationships between the surface and the ambient space; the first and second forms are here conditioned, as is seen, by the third form.

The relationships which we are going to find between the coefficients of the first and of the second forms, are due to the fact that already in the definition of the first form, the third has been taken into account.

The relationships referred to are the following

$$E = D^2 + \frac{D'^2}{\cos^2\varphi}; \qquad F = D'\left(D + \frac{D''}{\cos^2\varphi}\right); \qquad G = D'^2 + \frac{D''^2}{\cos^2\varphi} \tag{17.1}$$

and then

$$\left.\begin{aligned} \Delta = DD'' - D'^2 = \sqrt{EG - F^2}\cos\varphi = \frac{\cos^2\varphi}{\mathrm{K}} \\ \frac{F}{E\cos^2\varphi - G} = \frac{D'}{D\cos^2\varphi - D''} \end{aligned}\right\} \tag{17.2}$$

can be immediately deduced.

After this, the expression of D, D', D'' by the second derivatives of the potential is straightforward; we have

$$D = -\frac{w_{22}}{\mathrm{K}}; \qquad D' = \frac{w_{12}\cos\varphi}{\mathrm{K}}; \qquad D'' = -\frac{w_{11}\cos^2\varphi}{\mathrm{K}}. \tag{17.3}$$

18. The Potential as Third Intrinsic Coordinate; Null Direction of σ

Passing now to the examination of the homography σ considered in the whole of space, and not only on Σ, we shall find it convenient to introduce a third coordinate which will allow us to distinguish one level surface from another. The coordinate which spontaneously springs to mind as the most apt for the purpose is the value W^* of the potential which holds for that surface; or what is the same, the value of its difference

$$W = W^* - W_0 \tag{18.1}$$

from the value W_0 of the potential proper to a surface taken as a reference (which in general will be the geoid defined as coincident with the mean level of the seas, where these exist). The value $W = W^* - W_0$ measures, at least to an inessential factor, the dynamic height; and represents the work necessary to transport a unit of mass from the surface of the geoid to the one considered.

Given this, we shall attribute to the infinitesimal displacement dP the form

$$dP = i_1'\sqrt{E}\,d\varphi + i_2'\sqrt{G}\,d\lambda + i_3'\sqrt{N}\,dW, \tag{18.2}$$

where i_3' will be the third base versor.

From the fundamental relationship

$$dn = \sigma dP = i_1 d\varphi + i_2 \cos\varphi\, d\lambda \tag{18.3}$$

it is seen that i_3' necessarily represents the null direction of σ, such that $\sigma i_3' = 0$. As has been seen in § 6, the null direction of σ is that of the vector

$$(K\sigma c_1) \wedge (K\sigma c_2) \tag{18.4}$$

and hence, with p as a coefficient of proportionality

$$i_3' = p\{(w_{12}w_{23} - w_{22}w_{13})i_1 + (w_{12}w_{13} - w_{11}w_{23})i_2 + (w_{11}w_{22} - w_{12}^2)i_2)\}. \tag{18.5}$$

From the relationship

$$\operatorname{grad} W \times \frac{\partial P}{\partial W}\, dW = -g i_3 \times i_3' \sqrt{N}\, dW = dW \tag{18.6}$$

we then have

$$p = \frac{-1}{g\mathrm{K}\sqrt{N}}; \tag{18.7}$$

and given

$$i_3'\sqrt{N} = -H_1 i_1 - H_2 i_2 - \frac{i_3}{g} \tag{18.8}$$

it follows that

$$H_1 = \frac{w_{12} w_{23} - w_{22} w_{13}}{g\mathrm{K}}; \qquad H_2 = \frac{w_{12} w_{13} - w_{11} w_{23}}{g\mathrm{K}} \tag{18.9}$$

$$N = H_1^2 + H_2^2 + \frac{1}{g^2}. \tag{18.10}$$

19. Fundamental Metric Homography

Let us denote by $\boldsymbol{v}_1, \boldsymbol{v}_2, \boldsymbol{v}_3$ the vectors

$$\boldsymbol{v}_1 = \frac{\partial P}{\partial \varphi} = \boldsymbol{i}'_1 \sqrt{E}; \qquad \boldsymbol{v}_2 = \frac{\partial P}{\partial \lambda} = \boldsymbol{i}'_2 \sqrt{G}; \qquad \boldsymbol{v}_3 = \frac{\partial P}{\partial W} = \boldsymbol{i}'_3 \sqrt{N}; \tag{19.1}$$

which form what we shall call the geodetic triad, not necessarily either unit, or orthogonal, which defines the fundamental metric homography μ when it is referred to the intrinsic coordinates φ, λ and W introduced here.

Let us consider first of all the homography ν which transforms $(\boldsymbol{i}_1, \boldsymbol{i}_2, \boldsymbol{i}_3)$ into $(\boldsymbol{v}_1, \boldsymbol{v}_2, \boldsymbol{v}_3)$, a homography which is written, in the local astronomical system of versors,

$$\nu = \begin{Bmatrix} \boldsymbol{v}_1 & \boldsymbol{v}_2 & \boldsymbol{v}_3 \\ \boldsymbol{i}_1 & \boldsymbol{i}_2 & \boldsymbol{i}_3 \end{Bmatrix} \tag{19.2}$$

and hence the homography $K\nu \cdot \nu$; we have

$$K\nu \cdot \nu \boldsymbol{i}_r \times \boldsymbol{i}_s = \nu \boldsymbol{i}_r \times \nu \boldsymbol{i}_s = \boldsymbol{v}_r \times \boldsymbol{v}_s. \tag{19.3}$$

The components of $K\nu \cdot \nu$, which will be called the fundamental metric homography, in the system $(\boldsymbol{i}_1, \boldsymbol{i}_2, \boldsymbol{i}_3)$, are thus the scalar products of the metric vectors $(\boldsymbol{v}_1, \boldsymbol{v}_2, \boldsymbol{v}_3)$ of the chosen intrinsic system. A knowledge of it allows us to calculate immediately the scalar product (and thus the modulus) of vectors whose components in the $(\boldsymbol{v}_1, \boldsymbol{v}_2, \boldsymbol{v}_3)$ system are known.

A characteristic property of the geodetic triad is given by the following fundamental relationships:

$$dP = \frac{\partial P}{\partial \varphi} d\varphi + \frac{\partial P}{\partial \lambda} d\lambda + \frac{\partial P}{\partial W} dW = \boldsymbol{v}_1 d\varphi + \boldsymbol{v}_2 d\lambda + \boldsymbol{v}_3 dW. \tag{19.4}$$

The components of the homography ν in the local astronomical system are given by $\boldsymbol{v}_r \times \boldsymbol{i}_s$, and thus by the matrix

$$\nu = \begin{Vmatrix} -D & -\dfrac{D'}{\cos\varphi} & 0 \\ -D' & -\dfrac{D''}{\cos\varphi} & 0 \\ -H_1 & -H_2 & -\dfrac{1}{g} \end{Vmatrix}. \tag{19.5}$$

On the other hand, the components of the homography $K\nu \cdot \nu$ are given by $\boldsymbol{v}_r \times \boldsymbol{v}_s$, as we have seen. We shall introduce the symbol

$$g_{rs} = \boldsymbol{v}_s \times \boldsymbol{v}_r = g_{sr} \tag{19.6}$$

and thus write

$$K\nu \cdot \nu = g_{rs} = \begin{Vmatrix} E & F & L \\ F & G & M \\ L & M & N \end{Vmatrix}. \tag{19.7}$$

The expressions for $E, F, G,$ and N have already been given; we have moreover

$$\left.\begin{aligned} L &= \boldsymbol{v}_1 \times \boldsymbol{v}_2 = \sqrt{EN} \boldsymbol{i}'_1 \times \boldsymbol{i}'_3 = DH_1 + \frac{D'H_2}{\cos\varphi} \\ M &= \boldsymbol{v}_2 \times \boldsymbol{v}_3 = \sqrt{GN} \boldsymbol{i}'_2 \times \boldsymbol{i}'_3 = D'H_1 + \frac{D''H_2}{\cos\varphi} \end{aligned}\right\} \tag{19.8}$$

and hence, in synthesis

$$\| g_{rs} \| = \begin{Vmatrix} D^2 + \dfrac{D'^2}{\cos^2\varphi} & D'\left(D + \dfrac{D''}{\cos^2\varphi}\right) & DH_1 + \dfrac{D'H_2}{\cos\varphi} \\ D'\left(D + \dfrac{D''}{\cos^2\varphi}\right) & D'^2 + \dfrac{D''^2}{\cos^2\varphi} & D'H_1 + \dfrac{D''H_2}{\cos\varphi} \\ DH_1 + \dfrac{D'H_2}{\cos\varphi} & D'H_1 + \dfrac{D''H_2}{\cos\varphi} & H_1^2 + H_2^2 + \dfrac{1}{g^2} \end{Vmatrix}. \tag{19.9}$$

The value D^2 of the determinant

$$\mathrm{D}^2 = | g_{rs} | \tag{19.10}$$

is given by

$$\mathrm{D}^2 = \frac{\Delta^2}{g^2 \cos^2\varphi} \tag{19.11}$$

since

$$\mathrm{D} = \boldsymbol{v}_1 \wedge \boldsymbol{v}_2 \times \boldsymbol{v}_3 = \frac{-\Delta}{g\cos\varphi}. \tag{19.12}$$

20. Isozenithal Lines; Meridians and Parallels

The direction of the versor $\boldsymbol{i}'_3$ will be called the isozenithal (or isovertical), whence $\sigma(\boldsymbol{i}'_3) = 0$. Denoting by ζ the zenith distance of the isozenithal versor, and by θ the azimuth of the vertical plane that contains it, we have

$$\operatorname{tg}\theta = \frac{H_2}{H_1}; \qquad \operatorname{tg}\zeta = g\sqrt{H_1^2 + H_2^2}. \tag{20.1}$$

In a quite analogous manner, we would have for the azimuths μ and π of the geodetic meridian and parallel

$$\operatorname{tg}\mu = \frac{-D'}{D\cos\varphi}; \qquad \operatorname{tg}\pi = \frac{-D''}{D'\cos\varphi}; \tag{20.2}$$

and we have furthermore the following equations, gathered here for the convenience of the reader

$$\left.\begin{aligned}
&H_1 = \frac{1}{g}\left(D w_{13} + \frac{D'}{\cos\varphi} w_{23}\right); \quad H_2 = \frac{1}{g\cos\varphi}\left(D' w_{13} + \frac{D''}{\cos\varphi} w_{23}\right);\\
&w_{13} = \frac{g}{\Delta}(D''H_1 - D'H_2\cos\varphi); \quad w_{23} = \frac{g\cos\varphi}{\Delta}(DH_2\cos\varphi - D'H_1);\\
&D = -\frac{w_{22}}{K}; \quad D' = \frac{w_{12}\cos\varphi}{K}; \quad D'' = -\frac{w_{11}\cos^2\varphi}{K}:\\
&\Delta = \frac{\cos^2\varphi}{K} = DD'' - D'^2;\\
&w_{11} = -\frac{D''}{\Delta}; \quad w_{12} = \frac{D'\cos\varphi}{\Delta}; \quad w_{22} = -\frac{D\cos^2\varphi}{\Delta}
\end{aligned}\right\}. \tag{20.3}$$

In what follows we shall have need to introduce two more new quantities, h and k, related to the foregoing by

$$\left.\begin{aligned}
&h = \frac{D''H_1}{\cos\varphi} - D'H_2 = \frac{w_{13}\Delta}{g\cos\varphi}; \quad k = DH_2 - \frac{D'H_1}{\cos\varphi} = \frac{w_{23}\Delta}{g\cos^2\varphi};\\
&H_1 = \frac{(Dh + D'k)\cos\varphi}{\Delta}; \quad H_2 = \frac{D'h + D''k}{\Delta}
\end{aligned}\right\}. \tag{20.4}$$

21. The Reciprocal Vectors

Alongside the fundamental geodetic vectors

$$v_1 = \frac{\partial P}{\partial\varphi}; \quad v_2 = \frac{\partial P}{\partial\lambda}; \quad v_3 = \frac{\partial P}{\partial W}, \tag{21.1}$$

which we shall also write from now on as

$$v_r = \frac{\partial P}{\partial y^r} \tag{21.2}$$

in which we take $y^1 = \varphi$, $y^2 = \lambda$, $y^3 = W$, it will often be convenient to consider a new triad of fundamental vectors, which we shall call reciprocal to the first, and which we shall denote by (v^1, v^2, v^3) such that

$$\left.\begin{aligned}
&v^1 \times v_1 = v^2 \times v_2 = v^3 \times v_3 = 1\\
&v^1 \times v_2 = v^2 \times v_1 = v^1 \times v_3 = v^3 \times v_1 = v^2 \times v_3 = v^3 \times v_2 = 0
\end{aligned}\right\}. \tag{21.3}$$

Equations (21.3) can be synthesized in the abbreviated notation

$$v^r \times v_s = \delta^r_s \tag{21.4}$$

in which δ_s^r is Kronecker's symbol, equal to unity for $r = s$, and equal to zero for $r \neq s$.

It is easy to see that the vectors of the reciprocal triad, also called the reciprocal vectors, are given by

$$\left.\begin{aligned} &v^1 = \frac{v_2 \wedge v_3}{\mathrm{D}}; \qquad v^2 = \frac{v_3 \wedge v_1}{\mathrm{D}}; \qquad v^3 = \frac{v_1 \wedge v_2}{\mathrm{D}} \\ &\mathrm{D} = v_1 \wedge v_2 \times v_3 = -\frac{\Delta}{g \cos\varphi} \end{aligned}\right\}. \tag{21.5}$$

The relationship of reciprocity is naturally invertible; we have

$$\left.\begin{aligned} &v_1 = \frac{v^2 \wedge v^3}{\mathrm{D}'}; \qquad v_2 = \frac{v^3 \wedge v^1}{\mathrm{D}'}; \qquad v_3 = \frac{v^1 \wedge v^2}{\mathrm{D}'} \\ &\mathrm{D}' = v^1 \wedge v^2 \times v^3 = \frac{1}{\mathrm{D}} = -\frac{g \cos\varphi}{\Delta} \end{aligned}\right\}. \tag{21.6}$$

It is easy to see now that rot $v^r = 0$; in fact we have

$$\operatorname{rot} v^r \times v^s = \frac{1}{\mathrm{D}} \operatorname{rot} v^r \times v_{s+1} \wedge v_{s+2} = \frac{1}{\mathrm{D}} \left(\frac{\partial v^r}{\partial y^{s+1}} \times v_{s+2} - \frac{\partial v^r}{\partial y^{s+2}} \times v_{s+1} \right) \tag{21.7}$$

from the definition itself of the operator rot, and putting by convention equal to each other all indices which differ by multiples of 3.

Now differentiating with respect to y^j the equation

$$v^r \times v_i = \delta_i^r \tag{21.8}$$

we then have

$$\frac{\partial v^r}{\partial y^j} \times v_i = -\frac{\partial v_i}{\partial y^j} \times v^r = -\frac{\partial v_j}{\partial y^i} \times v^r = \frac{\partial v^r}{\partial y^i} \times v_j \tag{21.9}$$

whence

$$\frac{\partial v_i}{\partial y^j} = \frac{\partial^2 P}{\partial y^i \partial y^j} = \frac{\partial v_j}{\partial y^i} \tag{21.10}$$

and thus

$$\operatorname{rot} v^r \times v^s = 0 ; \qquad \operatorname{rot} v^r = 0 . \tag{21.11}$$

In consequence, we have

$$v^r = \operatorname{grad} f^r \tag{21.12}$$

where by f^r is indicated a scalar function, whose significance will become apparent shortly. From the relationships

$$v^r \times v_s = \operatorname{grad} f^r \times \frac{\partial P}{\partial y^s} = \delta_s^r \tag{21.13}$$

we have immediately

$$f^r = y^r + \text{const}\,. \tag{21.14}$$

The reciprocal vectors are thus the gradients of the latitude, longitude, and potential, respectively.

Summarizing, we have the following group of fundamental formulae

$$\boldsymbol{v}_r = \frac{\partial P}{\partial y^r}; \qquad \boldsymbol{v}^r = \operatorname{grad} y^r \tag{21.15}$$

$$dP = \sum_{r=1}^{3} \boldsymbol{v}_r dy^r; \qquad dy^r = \boldsymbol{v}^r \times dP\,. \tag{21.16}$$

22. Components of the Infinitesimal Displacement

Recalling now the expressions (16.14), (18.9) and (21.3), it is easy to write the expression of the fundamental vectors and of the reciprocal vectors using the local astronomical versors. In the following we set out the six groups of vector equations which relate amongst themselves the three systems of vectors considered up till now, and which permit an easy passage from one reference to the other.

$$\left.\begin{aligned} \boldsymbol{v}_1 &= -D\boldsymbol{i}_1 - \frac{D'}{\cos\varphi}\boldsymbol{i}_2 \\ \boldsymbol{v}_2 &= -D'\boldsymbol{i}_1 - \frac{D''}{\cos\varphi}\boldsymbol{i}_2 \\ \boldsymbol{v}_3 &= -H_1\boldsymbol{i}_1 - H_2\boldsymbol{i}_2 - \frac{1}{g}\boldsymbol{i}_3 \end{aligned}\right\} \tag{22.1}$$

$$\left.\begin{aligned} \boldsymbol{i}_1 &= -\frac{D''}{\Delta}\boldsymbol{v}_1 + \frac{D'}{\Delta}\boldsymbol{v}_2 \\ \boldsymbol{i}_2 &= \frac{D'\cos\varphi}{\Delta}\boldsymbol{v}_1 - \frac{D\cos\varphi}{\Delta}\boldsymbol{v}_2 \\ \boldsymbol{i}_3 &= \frac{gh\cos\varphi}{\Delta}\boldsymbol{v}_1 + \frac{gk\cos\varphi}{\Delta}\boldsymbol{v}_2 - g\boldsymbol{v}_3 \end{aligned}\right\} \tag{22.2}$$

$$\left.\begin{aligned} \boldsymbol{v}^1 &= -\frac{D''}{\Delta}\boldsymbol{i}_1 + \frac{D'\cos\varphi}{\Delta}\boldsymbol{i}_2 + \frac{gh\cos\varphi}{\Delta}\boldsymbol{i}_3 \\ \boldsymbol{v}^2 &= \frac{D'}{\Delta}\boldsymbol{i}_1 - \frac{D\cos\varphi}{\Delta}\boldsymbol{i}_2 + \frac{gk\cos\varphi}{\Delta}\boldsymbol{i}_3 \\ \boldsymbol{v}^3 &= -g\boldsymbol{i}_3 \end{aligned}\right\} \tag{22.3}$$

$$\left.\begin{aligned} \boldsymbol{i}_1 &= -D\boldsymbol{v}^1 - D'\boldsymbol{v}^2 - H_1\boldsymbol{v}^3 \\ \boldsymbol{i}_2 &= -\frac{D'}{\cos\varphi}\boldsymbol{v}^1 - \frac{D''}{\cos\varphi}\boldsymbol{v}^2 - H_2\boldsymbol{v}^3 \\ \boldsymbol{i}_3 &= -\frac{1}{g}\boldsymbol{v}^3 \end{aligned}\right\} \tag{22.4}$$

$$\left.\begin{aligned} v_1 &= E v^1 + F v^2 + L v^3 = \sum_r g_{1r} v^r \\ v_2 &= F v^1 + G v^2 + M v^3 = \sum_r g_{2r} v^r \\ v_3 &= L v^1 + M v^2 + N v^3 = \sum_r g_{3r} v^r \end{aligned}\right\} \tag{22.5}$$

$$\left.\begin{aligned} v^1 &= \frac{\cos^2\varphi}{\Delta^2}\{(G+g^2h^2)v_1 - (F-g^2hk)v_2\} - \frac{g^2\cos\varphi}{\Delta} h v_3 = \sum_r g^{1r} v_r \\ v^2 &= \frac{\cos^2\varphi}{\Delta^2}\{-(F-g^2hk)v_1 + (E+g^2k^2)v_2\} - \frac{g^2\cos\varphi}{\Delta} k v_3 = \sum_r g^{2r} v_r \\ v^3 &= -\frac{g^2\cos\varphi}{\Delta}(hv_1 + kv_2) + g^2 v_3 = \sum_r g^{3r} v_r \end{aligned}\right\} \tag{22.6}$$

where h and k are the quantities already defined in Eq. (20.4), and where g^{rs} denotes the scalar products

$$g^{rs} = v^r \times v^s = g^{sr}. \tag{22.7}$$

We note also that the matrix of the g^{rs} is the reciprocal of the g_{rs}, and vice versa.

In a quite analogous manner, it is now easy to find the relationships which connect the components (dx^1, dx^2, dx^3), $(d\varphi, d\lambda, dW)$, $(\overline{d\varphi}, \overline{d\lambda}, \overline{dW})$ of the displacement dP, in the three systems considered, $(\boldsymbol{i}_r)$, $(\boldsymbol{v}_r)$ and $(\boldsymbol{v}^r)$ respectively:

$$\left.\begin{aligned} dy_1 &= -D\,dx^1 - \frac{D'}{\cos\varphi} dx^2 \\ dy_2 &= -D'dx^1 - \frac{D''}{\cos\varphi} dx^2 \\ dy_3 &= -H_1 dx^1 - H_2 dx^2 - \frac{1}{g} dx^3 \end{aligned}\right\} \tag{22.8}$$

$$\left.\begin{aligned} dx^1 &= -\frac{D''}{\Delta} dy_1 + \frac{D'}{\Delta} dy_2 \\ dx^2 &= \frac{D'\cos\varphi}{\Delta} dy_1 - \frac{D\cos\varphi}{\Delta} dy_2 \\ dx^3 &= \frac{gh\cos\varphi}{\Delta} dy_1 + \frac{gk\cos\varphi}{\Delta} dy_2 - g\,dW \end{aligned}\right\} \tag{22.9}$$

$$\left.\begin{aligned} dy^1 &= -\frac{D''}{\Delta} dx^1 + \frac{D'\cos\varphi}{\Delta} dx^2 + \frac{gh\cos\varphi}{\Delta} dx^3 \\ dy^2 &= \frac{D'}{\Delta} dx^1 - \frac{D\cos\varphi}{\Delta} dx^2 + \frac{gk\cos\varphi}{\Delta} dx^3 \\ dy^3 &= -g\,dx^3 \end{aligned}\right\} \tag{22.10}$$

$$\left.\begin{aligned} dx^1 &= -D dy' \quad - D' dy^2 - H_1 dy^3 \\ dx^2 &= -\frac{D'}{\cos\varphi} dy^1 - \frac{D''}{\cos\varphi} dy^2 - H_2 dy^3 \\ dx^3 &= -\frac{1}{g} dy^3 \end{aligned}\right\} \tag{22.11}$$

$$dy_r = \sum_s g_{rs} dy^s \tag{22.12}$$

$$dy^r = \sum_s g^{rs} dy_s \tag{22.13}$$

having written $\overline{d\varphi} = dy_1, \overline{d\lambda} = dy_2, \overline{dW} = dy_3$.

23. Controvariant and Covariant Components

If we represent the homography v in the local astronomical system, according to Eq. (19.5), denoting by a_r^λ the components

$$v \equiv \| a_r^\lambda \| \equiv \begin{Vmatrix} a_1^1 & a_1^2 & a_1^3 \\ a_2^1 & a_2^2 & a_2^3 \\ a_3^1 & a_3^2 & a_3^2 \end{Vmatrix} \tag{23.1}$$

and thus the homography

$$v^{-1} \equiv A_r^\lambda \equiv \begin{Vmatrix} -\dfrac{D''}{\Delta} & \dfrac{D'}{\Delta} & 0 \\ \dfrac{D' \cos\varphi}{\Delta} & -\dfrac{D \cos\varphi}{\Delta} & 0 \\ \dfrac{gh \cos\varphi}{\Delta} & \dfrac{gk \cos\varphi}{\Delta} & -g \end{Vmatrix} \tag{23.2}$$

then the relationships of § 22 are written rather more simply below (taking here as the sign of summation those indices which appear twice, once above and once below, and which are known as dummy indices)

$$\begin{aligned} &\left.\begin{aligned} \boldsymbol{v}_r &= a_r^\lambda \boldsymbol{i}_\lambda & \quad \boldsymbol{i}_r &= A_r^\lambda \boldsymbol{v}_\lambda \\ \boldsymbol{v}^r &= A_\lambda^r \boldsymbol{i}^\lambda & \quad \boldsymbol{i}^r &= a_\lambda^r \boldsymbol{v}^\lambda \\ \boldsymbol{v}_r &= g_{rs} \boldsymbol{v}^s & \quad \boldsymbol{v}^r &= g^{rs} \boldsymbol{v}_s \end{aligned}\right\} \\ &\left.\begin{aligned} dy_r &= a_r^\lambda dx_\lambda & \quad dx_r &= A_r^\lambda dy_\lambda \\ dy^r &= A_\lambda^r dx^\lambda & \quad dx^r &= a_\lambda^r dy^\lambda \\ dy_r &= g_{rs} dy^s & \quad dy^r &= g^{rs} dy_s \end{aligned}\right\} . \end{aligned} \tag{23.3}$$

For the fundamental astronomical versors $(\boldsymbol{i}_r)$ and for the components of displacement along these, the indices are placed indifferently above or below, according to the occasion. This is legitimate in this case, when we are dealing with an orthogonal system of versors, which coincide with their own reciprocals.

It follows from the equations written that

the dy_r are transformed like the vectors $\boldsymbol{v}_r$
the dy^r are transformed like the vectors $\boldsymbol{v}^r$

and hence it is said that

the dy_r are cogradient to the base vectors $\boldsymbol{v}_r$
the dy^r are cogradient to the base vectors $\boldsymbol{v}^r$
the dy_r are controgradient to the base vectors $\boldsymbol{v}^r$
the dy^r are controgradient to the base vectors $\boldsymbol{v}_r$.

Reserving a privileged position for the fundamental metric vectors $(\boldsymbol{v}_r)$ introduced first, whose geometric and mechanical significance is more immediate, we shall say that the dy_r are covariant, and the dy^r are controvariant with respect to such vectors. The first are contradistinguished, in the more general symbolism already introduced, by an index below, the second by an index above.

For any vector whatever, for a homography, or hyperhomography, we can likewise consider the components according to the fundamental metric triad $(\boldsymbol{v}_r)$, or according to the reciprocal triad $(\boldsymbol{v}^r)$; the first will therefore be called controvariant, the second covariant.

In the rest of the treatment, all the absolute entities considered up till now will be expressed by means of such components; these represent scalars accessible to direct or indirect measurement, by means of the operations of geometric Geodesy, of dynamic Geodesy, or of Geodetic Astronomy.

If we denote indifferently by T_λ or T^λ, $T_{\lambda\mu}$ or $T^{\lambda\mu}$, etc. the components of a vector or of a homography in the local astronomical system of versors, and by T_r, T^r, T_{rs}, etc. their covariant, controvariant, or also mixed components in the geodetic system, then

$$\left.\begin{array}{ll} T_r = a_r^\lambda T_\lambda & T_r = A_r^\lambda T_\lambda \\ T_{rs} = a_r^\lambda a_s^\mu T_{\lambda\mu} & T_{rs} = A_r^\lambda A_s^\mu T_{\lambda\mu} \\ T^r = A_\lambda^r T^\lambda & T^r = a_\lambda^r T^\lambda \\ T_{r\cdot}^{\cdot s} = a_r^\lambda A_\mu^s T_{\lambda\cdot}^{\cdot\mu} & T_{r\cdot}^{\cdot s} = A_r^\lambda a_\mu^s T_{\lambda\cdot}^{\cdot\mu} \end{array}\right\} . \tag{23.4}$$

Considered from this point of view, the entities under examination take the name of tensors; the vectors correspond precisely to simple tensors, the homographies to double tensors, and so on.

When such entities are given in absolute form, the determination of their components is immediate. For example, let us consider a vector $\boldsymbol{u}$ whose components in the local astronomical system are u_r $(= u^r$ for the reasons already given), whose covariant components in the geodetic system $(\boldsymbol{v}_r)$ are u_r, and controvariant u^r; we shall have

$$\boldsymbol{u} = u^r \boldsymbol{i}_r = u^r \boldsymbol{v}_r = u_r \boldsymbol{v}^r \tag{23.5}$$

and also

$$u^r = u_r = \boldsymbol{u} \times \boldsymbol{i}_r ; \qquad u_r = \boldsymbol{u} \times \boldsymbol{v}_r ; \qquad u^r = \boldsymbol{u} \times \boldsymbol{v}^r . \tag{23.6}$$

Similarly, for a homography τ

$$\left.\begin{array}{lll} T_{rs} = T^{rs} = T^{r\cdot}_{\cdot s} = \tau \boldsymbol{i}_r \times \boldsymbol{i}_s\,; & T_{rs} = \tau \boldsymbol{v}_r \times \boldsymbol{v}_s\,; & T^{rs} = \tau \boldsymbol{v}^r \times \boldsymbol{v}^s\,; \\ T^{\cdot s}_{r\cdot} = \tau \boldsymbol{v}_r \times \boldsymbol{v}^s\,; & T^{r\cdot}_{\cdot s} = \tau \boldsymbol{v}^r \times \boldsymbol{v}_s\,; & \text{etc.} \end{array}\right\}. \tag{23.7}$$

Let us consider in particular the gradient of a scalar f; this represents a vector whose components in the system $(\boldsymbol{i}_r)$ are given by

$$F_r = \frac{\partial f}{\partial x^r}\,. \tag{23.8}$$

In the $(\boldsymbol{v}_r)$ system we have on the other hand

$$F_r = a^\lambda_r F_\lambda = \frac{\partial f}{\partial y^r} \tag{23.9}$$

since

$$a^\lambda_r = \frac{\partial x^\lambda}{\partial y^r}\,. \tag{23.10}$$

The derivatives of a scalar with respect to the controvariant coordinates thus represent the covariant components of its gradient.

24. Absolute Differentiation; Covariant Differentiation

Given a vector $\boldsymbol{u}$ by means of its covariant components u_r:

$$\boldsymbol{u} = u_r \boldsymbol{v}^r \tag{24.1}$$

we have, differentiating

$$d\boldsymbol{u} = du_r \boldsymbol{v}^r + u_r d\boldsymbol{v}^r = \left(\frac{\partial u_k}{\partial y^k} \boldsymbol{v}^k + u_r \frac{\partial \boldsymbol{v}^r}{\partial y^h}\right) dy^h\,. \tag{24.2}$$

Expressing the $\dfrac{\partial \boldsymbol{v}^r}{\partial y^h}$ by the $\boldsymbol{v}^k$ themselves, we shall put as usual

$$\frac{\partial \boldsymbol{v}^r}{\partial y^h} = -\Gamma^r_{hk} \boldsymbol{v}^k \tag{24.3}$$

by which Eq. (24.2) becomes

$$d\boldsymbol{u} = \left(\frac{\partial u_k}{\partial y^h} - u_r \Gamma^r_{hk}\right) \boldsymbol{v}^k dy^h = u_{k/h} \boldsymbol{v}^k dy^h\,. \tag{24.4}$$

The expression

$$u_{k/h} = \frac{\partial u_k}{\partial y^h} - u_r \Gamma^r_{hk} \tag{24.5}$$

is the covariant (or tensor) derivative of u_r; and Eq. (24.4) shows that, with respect to tensor differentiation, the base vectors v^k behave as if they were constant. This is in essence the *lemma of Ricci*.

From Eq. (24.3) we then have, from scalar multiplication by v_k:

$$\Gamma^r_{hk} = -\frac{\partial v^r}{\partial y^h} \times v_k ; \tag{24.6}$$

where the systems of three indices Γ^r_{hk} are Christoffel's symbols of the second kind. Taking account that $v^r \times v_s = \delta^r_s$ we have immediately

$$\Gamma^r_{hk} = -\frac{\partial v^r}{\partial y^k} \times v_h = \frac{\partial v_h}{\partial y^k} \times v^r = \frac{\partial v_k}{\partial y^h} \times v^r = -\frac{\partial v^r}{\partial y^h} \times v_k = \Gamma^r_{kh} . \tag{24.7}$$

In a quite analogous manner we would have for the covariant derivative of a double tensor

$$T_{hk/_l} = \frac{\partial T_{hk}}{\partial y^l} - T_{hj}\Gamma^j_{kl} - T_{jk}\Gamma^j_{hl} . \tag{24.8}$$

25. The Conditions of Integrability

Let us observe first of all that by an easy computation we have

$$H_1 = \frac{\partial}{\partial \varphi}\left(\frac{1}{g}\right); \qquad H_2 \cos\varphi = \frac{\partial}{d\lambda}\left(\frac{1}{g}\right). \tag{25.1}$$

Given this, the calculation of the partial derivatives of the fundamental metric vectors straightaway yields

$$\left.\begin{aligned}
\frac{\partial v_1}{\partial y^1} &= -\frac{\partial D}{\partial \varphi} i_1 - \left(\frac{\partial D'}{\partial \varphi} + D' \operatorname{tg}\varphi\right)\frac{i_2}{\cos\varphi} + D i_3 \\
\frac{\partial v_1}{\partial y^2} &= -\left(\frac{\partial D}{\partial \lambda} + D' \operatorname{tg}\varphi\right) i_1 - \left(\frac{\partial D'}{\partial \lambda} - D \sin\varphi\cos\varphi\right)\frac{i_2}{\cos\varphi} + D' i_3 \\
\frac{\partial v_1}{\partial y^3} &= -\frac{\partial D}{\partial W} i_1 - \frac{\partial D'}{\partial W}\frac{i_2}{\cos\varphi}
\end{aligned}\right\} \tag{25.2}$$

$$\left.\begin{aligned}
\frac{\partial v_2}{\partial y^1} &= -\frac{\partial D'}{\partial \varphi} i_1 - \left(\frac{\partial D''}{\partial \varphi} + D'' \operatorname{tg}\varphi\right)\frac{i_2}{\cos\varphi} + D' i_3 \\
\frac{\partial v_2}{\partial y^2} &= -\left(\frac{\partial D'}{\partial \lambda} + D'' \operatorname{tg}\varphi\right) i_1 - \left(\frac{\partial D''}{\partial \lambda} - D' \sin\varphi\cos\varphi\right)\frac{i_2}{\cos\varphi} + D'' i_3 \\
\frac{\partial v_2}{\partial y^3} &= -\frac{\partial D'}{\partial W} i_1 - \frac{\partial D''}{\partial W}\frac{i_2}{\cos\varphi}
\end{aligned}\right\} \tag{25.3}$$

$$
\left.
\begin{aligned}
\frac{\partial \boldsymbol{v}_3}{\partial y^1} &= -\left(\frac{\partial H_1}{\partial \varphi} + \frac{1}{g}\right)\boldsymbol{i}_1 - \frac{\partial H_2}{\partial \varphi}\boldsymbol{i}_2 \\
\frac{\partial \boldsymbol{v}_3}{\partial y^2} &= -\left(\frac{\partial H_1}{\partial \lambda} + H_2 \sin\varphi\right)\boldsymbol{i}_1 - \left(\frac{\partial H_2}{\partial \lambda} - H_1 \sin\varphi + \frac{\cos\varphi}{g}\right)\boldsymbol{i}_2 \\
\frac{\partial \boldsymbol{v}_3}{\partial y^3} &= -\frac{\partial H_1}{\partial W}\boldsymbol{i}_1 - \frac{\partial H_2}{\partial W}\boldsymbol{i}_2 - \frac{\partial}{\partial W}\left(\frac{1}{g}\right)\boldsymbol{i}_3
\end{aligned}
\right\} \tag{25.4}
$$

in which it should be noted that the components of $\frac{\partial \boldsymbol{v}_3}{\partial y^1}$ and of $\frac{\partial \boldsymbol{v}_3}{\partial y^2}$ with respect to $\boldsymbol{i}_3$ are zero, in virtue of Eq. (25.1). The equality of the mixed derivatives of P

$$
\frac{\partial \boldsymbol{v}_1}{\partial y^2} = \frac{\partial \boldsymbol{v}_2}{\partial y^1}; \qquad \frac{\partial \boldsymbol{v}_1}{\partial y^3} = \frac{\partial \boldsymbol{v}_3}{\partial y^1}; \qquad \frac{\partial \boldsymbol{v}_2}{\partial y^3} = \frac{\partial \boldsymbol{v}_3}{\partial y^2} \tag{25.5}
$$

is translated into the conditions of integrability which the coefficients D, D', D'', H_1, H_2, and $1/g$ must satisfy to ensure the existence of the coordinate surfaces φ, λ, and W. Such conditions are known as the equations of Mainardi-Codazzi. They are written for the φ surfaces

$$
\left.
\begin{aligned}
&\frac{\partial H_1}{\partial \lambda} - \frac{\partial D'}{\partial W} + H_2 \sin\varphi = 0 \\
&\frac{\partial H_2}{\partial \lambda} - \frac{\partial D''}{\partial W}\frac{1}{\cos\varphi} - H_1 \sin\varphi + \frac{1}{g}\cos\varphi = 0
\end{aligned}
\right\}, \tag{25.6}
$$

for the λ surfaces

$$
\left.
\begin{aligned}
&\frac{\partial H_1}{\partial \varphi} - \frac{\partial D}{\partial W} + \frac{1}{g} = 0 \\
&\frac{\partial H_2}{\partial \varphi} - \frac{\partial D'}{\partial W}\frac{1}{\cos\varphi} = 0
\end{aligned}
\right\} \tag{25.7}
$$

and for the W surfaces

$$
\left.
\begin{aligned}
&\frac{\partial D}{\partial \lambda} - \frac{\partial D'}{\partial \varphi} + D' \operatorname{tg}\varphi = 0 \\
&\frac{\partial D''}{\partial \varphi} - \frac{\partial D'}{\partial \lambda} + D \sin\varphi\cos\varphi + D'' \operatorname{tg}\varphi = 0
\end{aligned}
\right\}. \tag{25.8}
$$

The last pair of equations are already to be found in a note by Mineo.

26. The Christoffel Symbols

After this, the actual calculation of the Christoffel symbols is very easy, on the basis of the defining formulae; we have in fact

$$\left.\begin{aligned}
\Gamma^1_{11} &= \frac{1}{\Delta}\left\{D''\frac{\partial D}{\partial\varphi} - 2D'\frac{\partial D'}{\partial\varphi} + D'\frac{\partial D}{\partial\lambda} + D\,g\,h\cos\varphi\right\}\\
\Gamma^1_{21} = \Gamma^1_{12} &= \frac{1}{\Delta}\left\{D''\frac{\partial D}{\partial\lambda} - D'\frac{\partial D''}{\partial\varphi} + D'\,g\,h\cos\varphi\right\}\\
\Gamma^1_{22} &= \frac{1}{\Delta}\left\{-D''\frac{\partial D''}{\partial\varphi} + 2D''\frac{\partial D'}{\partial\lambda} - D'\frac{\partial D''}{\partial\lambda} + D''\,g\,h\cos\varphi\right\}\\
&\quad - \sin\varphi\cos\varphi\\
\Gamma^1_{13} = \Gamma^1_{31} &= \frac{1}{\Delta}\left\{D''\frac{\partial D}{\partial W} - D'\frac{\partial D'}{\partial W}\right\}\\
\Gamma^1_{23} = \Gamma^1_{32} &= \frac{1}{\Delta}\left\{D''\frac{\partial D'}{\partial W} - D'\frac{\partial D''}{\partial W}\right\}\\
\Gamma^1_{33} &= \frac{1}{\Delta}\left\{D''\frac{\partial H_1}{\partial W} - D'\frac{\partial H_2}{\partial W}\cos\varphi - g\frac{\partial}{\partial W}\left(\frac{1}{g}\right)\cdot h\cos\varphi\right\}
\end{aligned}\right\} \qquad (26.1)$$

$$\left.\begin{aligned}
\Gamma^2_{11} &= \frac{1}{\Delta}\left\{-D'\frac{\partial D}{\partial\varphi} + 2D\frac{\partial D'}{\partial\varphi} - D\frac{\partial D}{\partial\lambda} + D\,g\,k\cos\varphi\right\}\\
\Gamma^2_{12} = \Gamma^2_{21} &= \frac{1}{\Delta}\left\{D\frac{\partial D''}{\partial\varphi} - D'\frac{\partial D}{\partial\lambda} + D'\,g\,k\cos\varphi\right\} + \operatorname{tg}\varphi\\
\Gamma^2_{22} &= \frac{1}{\Delta}\left\{D\frac{\partial D''}{\partial\varphi} - 2D'\frac{\partial D'}{\partial\lambda} + D\frac{\partial D''}{\partial\lambda} + D''\,g\,k\cos\varphi\right\}\\
\Gamma^2_{13} = \Gamma^2_{31} &= \frac{1}{\Delta}\left\{D\frac{\partial D'}{\partial W} - D'\frac{\partial D}{\partial W}\right\}\\
\Gamma^2_{23} = \Gamma^2_{32} &= \frac{1}{\Delta}\left\{D\frac{\partial D''}{\partial W} - D'\frac{\partial D'}{\partial W}\right\}\\
\Gamma^2_{33} &= \frac{1}{\Delta}\left\{D\frac{\partial H_2}{\partial W}\cos\varphi - D'\frac{\partial H_1}{\partial W} - g\frac{\partial}{\partial W}\left(\frac{1}{g}\right)\cdot k\cos\varphi\right\}
\end{aligned}\right\} \qquad (26.2)$$

$$\left.\begin{aligned}
&\Gamma^3_{11} = -gD; \qquad \Gamma^3_{12} = \Gamma^3_{21} = -gD'; \qquad \Gamma^3_{22} = -gD'';\\
&\Gamma^3_{13} = \Gamma^3_{31} = \Gamma^3_{23} = \Gamma^3_{32} = 0;\\
&\Gamma^3_{33} = -\frac{\partial}{\partial W}\lg g = g\frac{\cos\varphi}{\Delta}(H_1 h + H_2 k\cos\varphi) + \frac{w_{33}}{g}\\
&\qquad\quad = g\frac{\cos^2\varphi}{\Delta^2}(Dh^2 + 2D'hk + D''k^2) + \frac{w_{33}}{g}
\end{aligned}\right\}. \qquad (26.3)$$

We note that with an easy calculation we have

$$I_3 \mathscr{W} = \frac{g \cos\varphi}{\mathrm{D}} \left\{ \frac{g}{\mathrm{D}^2} (D''k^2 + 2D'hk + Dh^2) - W_{33} \right\} \tag{26.4}$$

whence

$$\Gamma_{33}^3 = -\frac{\partial}{\partial W} \lg g = \frac{\mathrm{D} I_3 \mathscr{W}}{g^2 \cos\varphi} \,. \tag{26.5}$$

From the defining formulae we also have immediately the following expressions for the Christoffel symbols of the second kind

$$\begin{aligned} \Gamma_{rs}^t &= \boldsymbol{v}^t \times \frac{\partial \boldsymbol{v}_r}{\partial y^s} = -\frac{\partial \boldsymbol{v}^t}{\partial y^s} \times \boldsymbol{v}_r = -\frac{\partial \operatorname{grad} y^t}{\partial y^s} \times \boldsymbol{v}_r \\ &= \boldsymbol{v}^t \times \frac{\partial \boldsymbol{v}_s}{\partial y^r} = -\frac{\partial \boldsymbol{v}^t}{\partial y^r} \times \boldsymbol{v}_s = -\frac{\partial \operatorname{grad} y^t}{\partial y^r} \times \boldsymbol{v}_s = \frac{\partial^2 P}{\partial y^r \partial y^s} \times \boldsymbol{v}^t \,. \end{aligned} \tag{26.6}$$

It follows from these relationships that the symbols Γ_{rs}^t are, without regard to sign, the covariant components of the homographies

$$\frac{d}{dP} \operatorname{grad} y^t = \frac{d\boldsymbol{v}^t}{dP} \tag{26.7}$$

of which that with $t = 3$ is the homography of Eötvös considered several times previously.

We note further that the Christoffel symbols are linked to D by the relationship

$$\Gamma_{\mu i}^{\mu} = \frac{\partial \lg \mathrm{D}}{\partial y^i} \,. \tag{26.8}$$

27. The Gravimetric Gradient; the Eötvös Tensor; the Tensor of Burali-Forti

The covariant components of $\boldsymbol{g} = \boldsymbol{v}^3$, the gravity vector, are obviously

$$\boldsymbol{g} \times \boldsymbol{v}_r = g_r = \delta_r^3 = (0, 0, 1) \,; \tag{27.1}$$

whereas the controvariant components are

$$\boldsymbol{g} \times \boldsymbol{v}^r = g^r = g^{3r} \,. \tag{27.2}$$

At the same time, the covariant components of the versor $\boldsymbol{n}$ normal to the equipotential surfaces are

$$\boldsymbol{n} \times \boldsymbol{v}_r = n_r = -\frac{1}{g} \delta_r^3 = \left(0, 0, -\frac{1}{g}\right) \tag{27.3}$$

and the controvariant ones

$$\boldsymbol{n} \times \boldsymbol{v}^r = n^r = -\frac{1}{g} g^{3r} \,. \tag{27.4}$$

The covariant components of the gravimetric gradient $\boldsymbol{G}$ are then

$$\left.\begin{aligned} g/_1 &= G_1 = \frac{\partial g}{\partial \varphi} = H_{13} = -g^2 H_1 = \boldsymbol{G} \times \boldsymbol{v}_1 \\ g/_2 &= G_2 = \frac{\partial g}{\partial \lambda} = H_{23} = -g^2 H_2 \cos\varphi = \boldsymbol{G} \times \boldsymbol{v}_2 \\ g/_3 &= G_3 = \frac{\partial g}{\partial W} = H_{33} = -g\Gamma^3_{33} = \boldsymbol{G} \times \boldsymbol{v}_3 \end{aligned}\right\} . \tag{27.5}$$

The covariant components of the homographies

$$\left.\begin{aligned} \mathscr{W} &= \frac{d\boldsymbol{g}}{dP}; \qquad \mathscr{H} = H(\boldsymbol{G}, \boldsymbol{g}); \\ \sigma &= \frac{d\boldsymbol{n}}{dP} = -\frac{1}{g}\mathscr{W} + \frac{1}{g^2}\mathscr{H} \end{aligned}\right\} \tag{27.6}$$

which we shall denote respectively by W_{rs}, H_{rs}, S_{rs} are obtained immediately.

We have already seen that

$$W_{rs} = -\Gamma^3_{rs} = g_{r/s} = \mathscr{W}\boldsymbol{v}_r \times \boldsymbol{v}_s \tag{27.7}$$

and hence

$$W_{rs} \equiv \left\| \begin{matrix} gD & gD' & 0 \\ gD' & gD'' & 0 \\ 0 & 0 & \dfrac{\partial}{\partial W} \lg g \end{matrix} \right\| . \tag{27.8}$$

The same expressions would be obtained naturally by transforming by Eq. (23.4) the components [Eq. (9.8)] of $\mathscr{W}$ in the local astronomical system.

In the same way we have

$$\mathrm{S}_{rs} = n_{r/s} = -\left(\frac{1}{g} g_r\right)_{/s} = -\frac{1}{g} g_{r/s} + \frac{1}{g^2} g_r g_{/s}$$

$$\mathrm{H}_{rs} \equiv \left\| \begin{matrix} 0 & 0 & -g^2 H_1 \\ 0 & 0 & -g^2 H_2 \cos\varphi \\ 0 & 0 & \dfrac{\partial g}{\partial W} \end{matrix} \right\| \tag{27.9}$$

$$\mathrm{S}_{rs} \equiv \left\| \begin{matrix} -D & -D' & -H_1 \\ -D' & -D'' & -H_2 \cos\varphi \\ 0 & 0 & 0 \end{matrix} \right\| . \tag{27.10}$$

28. General Considerations

The equations found for the expressions of the simple and double tensors, which represent covariantly the vectors and homographies under examination, now permit us to transport the properties of these entities from one point to another in space, by exploiting the methods of the absolute tensor or differential calculus of Ricci and Levi-Civita. To carry out such a transport, and the comparison with the equivalent entities relative to another point in space, the consideration of the Christoffel symbols of the second kind is of fundamental importance, since these allow us to perform the differentiation of the fundamental metric vectors. We have directly

$$\left.\begin{aligned}
&\frac{\partial P}{\partial y^i} = v_i; \qquad \frac{\partial^2 P}{\partial y^i \partial y^j} = \frac{\partial v_i}{\partial y^j} = \frac{\partial v_j}{\partial y^i} = \Gamma^{\mu}_{ij} v_\mu; \\
&\frac{\partial^3 P}{\partial y^i \partial y^j \partial y^k} = \left(\frac{\partial \Gamma^{\mu}_{ij}}{\partial y^k} + \Gamma^{\lambda}_{ij}\Gamma^{\mu}_{\lambda k}\right) v_\mu; \\
&\frac{\partial^4 P}{\partial y^i \partial y^j \lambda y^k \partial y^l} = \left(\frac{\partial^2 \Gamma^{\mu}_{ij}}{\partial y^k \partial y^l} + \Gamma^{\mu}_{\lambda k}\frac{\partial \Gamma^{\lambda}_{ij}}{\partial y^l} + \Gamma^{\lambda}_{ij}\frac{\partial \Gamma^{\mu}_{\lambda k}}{\partial y^l}\right. \\
&\qquad\qquad \left. + \Gamma^{\mu}_{\nu l}\frac{\partial \Gamma^{\nu}_{ij}}{\partial y^k} + \Gamma^{\lambda}_{ij}\Gamma^{\mu}_{\nu l}\Gamma^{\nu}_{\lambda k}\right) v_\mu
\end{aligned}\right\} . \tag{28.1}$$

These relationships, and the others which we would be able to deduce by further differentiation of the foregoing, essentially translate the link existing between one geodetic triad, and the others of the field; from this viewpoint the Christoffel symbols of the second kind act as *connection coefficients* between one triad and the other. Covariant differentiation translates the same fact: the first term, comprising the partial derivative of the components of the tensor, represents the rate of variation of these as if the axes were invariant from point to point; the second term, expressed by Christoffel symbols, translates the contribution due, in the differentiation, to the fact that the fundamental triad changes as it passes from one point to another infinitesimally close.

Having rejected a single system of Cartesian axes, as being inaccessible to direct observation, it has now become clear how we might be enabled to procure in the way described an intrinsic coordinate system, comprising a triple infinity of coordinate triads in space, changing from point to point, but whose connection coefficients (or Christoffel symbols) permit us to deduce their mutual positions.

For each entity, whether vectorial or homographic, it will thus be possible to study the components, firstly with respect to a relative triad at the point P, chosen in relation to the nature of the problem under examination, and then to find the components of the same entity with respect to another triad, relative to another point P' in space, where it will then be possible to carry out comparisons with homologous entities.

D Applications

29. Premise

In this chapter dedicated to the foundations of the theory, there is not the slightest pretext for dealing with the applications which concern all the fields of

Geodesy, Geodetic Astronomy, and Gravimetry. However, one should not altogether avoid casting a fleeting glance at some immediate applications, above all with the aim of showing the simplicity with which the theory set out can be adapted in practice.

One of the applications, the first, concerns both the absolute and the local astronomical approaches, and deals in fact with the general theory of all the possible types of torsion balance.

The other applications concern instead the general (intrinsic) geodetic approach, and deal above all with the development in series of several scalar or vector entities. Such series developments are of capital importance, since they allow us to perform the numerical calculation of the Christoffel symbols from the observational data, and also of the covariant components of the various vectors and tensors which are involved.

30. Torsion Balances of Various Types

The most general type of torsion balance can be synthesized by a balance-arm suspended from its barycentre, carrying two equal masses at its ends. If the vector $\boldsymbol{a}$ represents in direction and magnitude the balance arm, the couple acting on it is then characterized by $\mathscr{W}\boldsymbol{a}$ and $\boldsymbol{a}$; and its vector moment by $(\mathscr{W}\boldsymbol{a}) \wedge \boldsymbol{a}$.

Taking $\boldsymbol{a}$ as a unit vector, then if we put

$$\boldsymbol{a} = \sin v \cos \alpha \cdot \boldsymbol{i}_1 + \sin v \sin \alpha \cdot \boldsymbol{i}_2 + \cos v \cdot \boldsymbol{i}_3 \tag{30.1}$$

we should have

$$\left.\begin{aligned}
&(\mathscr{W}\boldsymbol{a}) \wedge \boldsymbol{a} \\
&= \left\{\frac{1}{2}(W_{33} - W_{22}) \sin 2v \sin \alpha + W_{23}(\sin^2 v \sin^2 \alpha - \cos^2 v)\right. \\
&\quad \left. + \frac{1}{2}(W_{13} \sin^2 v \sin 2\alpha - W_{12} \sin 2v \cos \alpha)\right\} \boldsymbol{i}_1 \\
&+ \left\{\frac{1}{2}(W_{11} - W_{33}) \sin 2v \cos \alpha + W_{13}(\cos^2 v - \sin^2 v \cos^2 \alpha)\right. \\
&\quad \left. + \frac{1}{2}(W_{12} \sin 2v \sin^2 \alpha - W_{23} \sin^2 v \sin 2\alpha)\right\} \boldsymbol{i}_2 \\
&+ \left\{\frac{1}{2}(W_{22} - W_{11}) \sin^2 v \sin 2\alpha + W_{12} \sin^2 v \cos 2\alpha\right. \\
&\quad \left. + \frac{1}{2} \sin 2v (W_{23} \cos \alpha - W_{13} \sin \alpha)\right\} \boldsymbol{i}_3
\end{aligned}\right\} . \tag{30.2}$$

As is seen, the second derivatives of two equal indices cannot be separated, in so far as one uses instruments based on the measurement of moments, as are the torsion balances; for these always appear coupled with the same coefficient. If, however, we take note of the equation of Laplace

$$W_{11} + W_{22} + W_{33} = 2\,\omega^2 \tag{30.3}$$

the terms can be split up, and all the derivatives determined separately.

In practice it is rather difficult to measure a vector moment; but it is possible to determine the scalar moment with respect to one axis. If this axis is vertical, we have the well-known torsion balances of Eötvös of the first and second kinds:

balance of the first kind: $\nu = 90°$ (balance-arm horizontal)

$$(\mathscr{W}\boldsymbol{a})\wedge\boldsymbol{a}\times\boldsymbol{i}_3 = \frac{1}{2}(W_{22} - W_{11})\sin 2\,\alpha + W_{12}\cos 2\,\alpha\,; \tag{30.4}$$

balance of the second kind: $\nu \neq 90°$ (balance-arm oblique)

$$(\mathscr{W}\boldsymbol{a})\wedge\boldsymbol{a}\times\boldsymbol{i}_3 = \frac{1}{2}(W_{22} - W_{11})\sin^2\nu\sin 2\,\alpha$$

$$+ W_{12}\sin^2\nu\cos 2\,\alpha + \frac{1}{2}\sin 2\,\nu(W_{23}\cos\alpha - W_{13}\sin\alpha)\,. \tag{30.5}$$

Berroth has also conceived a balance oscillating about a horizontal axis normal to the vertical plane which contains the balance-arm. In this case we have

$$(\mathscr{W}\boldsymbol{a})\wedge\boldsymbol{a}\times(-\sin\alpha\cdot\boldsymbol{i}_1 + \cos\alpha\cdot\boldsymbol{i}_2)$$

$$= \frac{1}{2}\sin 2\,\nu[(W_{11} - W_{33})\cos^2\alpha + (W_{22} - W_{33})\sin^2\alpha + W_{12}\sin 2\,\alpha]$$

$$+ \cos 2\,\nu(W_{13}\cos\alpha + W_{23}\sin\alpha)\,. \tag{30.6}$$

It is also possible to conceive of a still more general balance, in which we might be able to measure at the same time the moment with respect to a vertical axis, and to a horizontal axis normal to the vertical plane containing the balance arm; such a balance would comprise a combination of that of Eötvös and that of Berroth.

31. Transport of the Gravity Vector and of the Gravimetric Gradient

The transport from a point P_0 to a point $P = P_0 + \Delta P$ at a finite distance, of the modulus g of gravity, comes about with the usual Taylor expansion

$$g - g_0 = \frac{\partial g}{\partial y^i}\Delta y^i + \frac{1}{2!}\,\frac{\partial^2 g}{\partial y^i \partial y^j}\Delta y^j \Delta y^i + \ldots\,. \tag{31.1}$$

If, however, we are dealing with entities of tensor character, we need to have regard, in the transport, to the system of base vectors to which such entities are taken to refer. Thus, for example, if we take the gravity vector $\boldsymbol{g}$, its components g_r are always $\delta_r^3 = (0,0,1)$ when referred to the local system of base vectors. If we wish, however, to compare the vectors $\boldsymbol{g}$ relative to two distinct points, for example P_0 and P_1, we shall need to refer both vectors to the same base triad, for example to that relative to P_0.

Thus in the first case the Taylor expansion of the ordinary derivatives of the components g_r will apply; and these are obviously all zero. In the second case we shall apply the Taylor expansion, but now for the tensor derivatives of g_r

$$g_r-(g_r)_0=g_{r/i}\Delta y^i+\frac{1}{2!}g_{r/ij}\Delta y^i\Delta y^j+\dots \tag{31.2}$$

recalling that $g_{r/i}=-\Gamma^3_{ri}$, and then

$$g_{r/ij}=-\frac{\partial\Gamma^3_{ri}}{\partial y^j}+2\Gamma^3_{jh}\Gamma^h_{ir}\quad\text{etc.} \tag{31.3}$$

In the same way, we would have for the gravimetric gradient

$$g_{/r}-(g_{/r})_0=g_{r/i}\Delta y^i+\frac{1}{2!}g_{/rij}\Delta y_j\Delta y_i+\dots \tag{31.4}$$

where

$$g_{/ri}=\frac{\partial H_{r3}}{\partial y^i}=H_{s3}\Gamma^s_{ri}\quad\text{etc.} \tag{31.5}$$

Quite similar equations are obtained for the transport for finite distances of the tensors of Eötvös and of Burali-Forti.

32. The Transport of Intrinsic Coordinates Along Space Curves

Given a curve of length s in space by means of the controvariant components

$$\lambda^i=\boldsymbol{t}\times\boldsymbol{v}^i=\frac{dy^i}{ds};\qquad \beta^i=\boldsymbol{B}\times\boldsymbol{v}^i \tag{32.1}$$

of its tangent $\boldsymbol{t}$ and of its binormal $\boldsymbol{B}$ at $P_0(y^r_0)$, and by means of its first and second curvatures γ and τ respectively, expressed as functions of the arc s, we seek the coordinates of its other end P and the corresponding values at P of λ^i and β^i. We have

$$\left.\begin{aligned} y^r&=y^r_{(0)}+s\lambda^r_{(0)}+\frac{s^2}{2!}\,\frac{d\lambda^r_0}{ds}+\dots\\ \lambda^r&=\lambda^r_{(0)}+s\frac{d\lambda^r_0}{ds}+\dots\\ \beta^r&=\beta^r_{(0)}+s\frac{d\beta^r_0}{ds}+\dots \end{aligned}\right\} \tag{32.2}$$

where, by the generalized Frenet equations

$$\frac{d\lambda^r}{ds}=\gamma v^r-\lambda^i\lambda^j\Gamma^r_{ij};\qquad \frac{d\beta^r}{ds}=\tau v^r-\beta^i\lambda^j\Gamma^r_{ij}\qquad\text{etc.} \tag{32.3}$$

and

$$v^r=\boldsymbol{N}\times\boldsymbol{v}^r\qquad(\boldsymbol{N}=\boldsymbol{B}\wedge\boldsymbol{t})\,. \tag{32.4}$$

These expansions generalize those of Legendre for the transport of the geographical coordinates and of the azimuth along a surface curve, extending them to distinct equipotential surfaces. Their importance resides in the fact that the modern tendency in Geodesy is to measure distances no longer along surface geodesics, but along geodesics (straight lines) or curves in space, which is made possible by modern radioelectric and interferential means.

33. The Equation of the Geoid

The relationships established up to now allow us to readily expand in power series the various geometric or mechanical entities which we are accustomed to consider in Geodesy, whether they be scalars, vectors (tensors of the first order), or homographies (tensors of the second order). If we now apply the Taylor expansion to the point P we would have

$$P = P_0 + \frac{\partial P}{\partial y^i} \Delta y^i + \frac{1}{2!} \frac{\partial^2 P}{\partial y^i \partial y^j} \Delta y^i \Delta y^j + \dots \tag{33.1}$$

where the $\frac{\partial P}{\partial y^i}, \frac{\partial^2 P}{\partial y^i \partial y^j}, \dots$ are given by Eq. (28.1), and where it is understood that the derivatives are all calculated at the point P_0.

Projecting the foregoing onto the local astronomical axes $(\boldsymbol{i}^r)$ relative to the point P_0, we have immediately the Cartesian coordinates of P with respect to such a triad, expressed by the geographic coordinates and by the dynamic height

$$\begin{aligned} \Delta x^r &= \frac{\partial P}{\partial y^i} \times \boldsymbol{i}^r \Delta y^i + \frac{1}{2!} \frac{\partial^2 P}{\partial y^i \partial y^j} \times \boldsymbol{i}_0^r \Delta y^i \Delta y^j + \dots \\ &= a_i^r \Delta y^i + \frac{1}{2!} a_\mu^r \Gamma_{ij}^\mu \Delta y^i \Delta y^j + \dots . \end{aligned} \tag{33.2}$$

Restricting in Eqs. (33.1) and (33.2) the summation on the indices i and j solely to the values 1 and 2, which is equivalent to setting $\Delta y^3 = 0$, we obviously obtain the equation of the level surface passing through P_0; if P_0 were situated at sea level, this would be the equation of the geoid. Equation (33.3) represents its most convenient parametric form

$$x^r = x^r(\varphi, \lambda) \tag{33.3}$$

as a function of the geographic coordinates.

Bibliography

I. General Treatises

Baeschlin CF (1947) Einführung in die Kurven- und Flächentheorie auf vektorieller Grundlage. Füssli, Zürich

Baeschlin CF (1948) Lehrbuch der Geodäsie. Füssli, Zürich

Bianchi L (1927) Lezioni di geometria differenziale. Zanichelli, Bologna

Brand L (1948) Vector and tensor analysis. Wiley, New York

Brillouin L (1947) Les tenseurs en mécanique et en élasticité. Masson, Paris
Burali-Forti C, Marcolongo R (1912) Analyse vectorielle générale. Mattei, Pavia
Burali-Forti C, Marcolongo R (1921) Elementi di calcolo vettoriale. Zanichelli, Bologna
Burali C, Marcolongo R (1929) Analisi vettoriale generale e applicazioni. Trasformazioni lineari. Zanichelli, Bologna
Burgatti P, Boggio T, Burali-Forti C (1930) Analisi vettoriale generale e applicazioni. Geometria differenziale. Zanichelli, Bologna
Cartan E (1946) Leçons sur la géométrie des espaces de Riemann. Gauthier-Villars, Paris
Darboux G (1889–1894) Leçons sur la théorie générale des surfaces et les applications du calcul infinitesimal. Gauthier-Villars, Paris
Eddington AS (1921) Espace, temps et gravitation; la théorie de la relativité généralisée dans ses grandes lignes. Hermann, Paris
Finzi B, Pastori M (1949) Calcolo tensoriale e applicazioni. Zanichelli, Bologna
Galbrun H (1923) Introduction géométrique à l'étude de la relativité. Gauthier-Villars, Paris
Gutenberg B (ed) (1929) Handbuch der Geophysik. Bornträger, Berlin; with later additional editions and supplements
Helmert FR (1880) Die mathematischen und physikalischen Theorien der höheren Geodäsie. Teubner, Leipzig. Reprint: Minerva GmbH, Frankfurt/M 1962
Hopfner F (1933) Physikalische Geodäsie. Akademische Verlagsgesellschaft, Leipzig
Hopfner F (1949) Grundlagen der höheren Geodäsie. Springer, Wien
Kellog OD (1929) Foundations of potential theory. Springer, Berlin
Laborde J (1945) La géométrie des sphéroides à l'usage des ingénieurs géographes. Inst Geogr Nat, Paris
Levi-Civita T (1925) Lezione di calcolo differenziale assoluto. Stock, Roma
Levi-Civita T (1947) The absolute differential calculus. Blackie, London Glasgow
Marais (1923) Introduction à la théorie de la relativité; calcul différentiel absolu et géométrie. Gauthier-Villars, Paris
Mc Connell AJ (1847) Applications of the absolute differential calculus. Blackie, London Glasgow
Pizzetti P (1913) Principi della teoria meccanica della figura dei pianeti. Spoerri, Pisa
Pizzetti P (1928) Trattato di geodesia teoretica. Zanichelli, Bologna
Poincaré H (1899) Théorie du potentiel Newtonien. Gauthier-Villars, Paris
Prey A, Mainka K, Tams E (1922) Einführung in die Geophysik. Julius Springer, Berlin
Ramos TA (1930) Leçons sur le calcul vectoriel. Blanchard, Paris
Ricci G, Levi-Civita T (1923) Méthodes de calcul différentiel absolu. Blanchard, Paris
Schouten JA (1924) Der Ricci-Kalkül. Springer, Berlin
Tardi P (1934) Traité de Géodésie. Gauthier-Villars, Paris
Wangerin A (1909) Theorie des Potentials und der Kugelfunktionen. Sammlung Schubert, Leipzig
Weyl H (1921) Raum, Zeit, Materie. Vorlesungen über allgemeine Relativitätstheorie. Springer, Berlin
Yuvet G (1922) Introduction au calcul tensoriel et au calcul différentiel absolu. Blanchard, Paris

Note – A full and systematic bibliography on tensor calculus is to be found in the "Calcolo tensoriale e applicazioni" (Tensor calculus and applications) by B. Finzi and M. Pastori, cited above.

II. Special Works

Ackerl F (1931) Das Geoid. Gerlands Beitr Geophys 29
Ackerl F (1932) Die Schwerkraft am Geoid. Oesterr Akad Sitzungsber Kl II Vol 14 l Vienna
Ballarin S (1935) Effetto dell'ellissoidicità della terra nella riduzione delle misure di gravità. Pubbl Ist Sup Ing Milan no 13
Benfield AE (1937) Note on the variation of gravity with depth. Z Geophys 13: no 4–5
Berroth A (1920) Eine vertikal schwingende Drehwaage. Z Instrumentenkd November 1920
Berroth A (1937) Gravitationsfeld und Oberflächengestalt der Erde und ihre geometrische Festlegung. Z Geophys 8
Boaga G (1941–42a) Il trasporto delle coordinate curvilinee lungo un arco di geodetica in alcuni casi particolari interessanti la Geodesia. Rend R Ist Lomb Sci Lett 75: fasc II

Boaga G (1941 – 42b) Teoria della bilancia di torsione. Universo December 1941, January – February 1942

Boaga G (1942) Sul trasporto delle coordinate curvilinee lungo un arco di geodetica su una superficie qualunque con applicazioni al problema geodetica del trasporto delle coordinate geografiche. Universo September 1942: no 7

Boriosi M (1932) Sviluppi in serie della gravità sulla superficie dell'ellissoide di rotazione. Boll Com Ital Geodes Geofis January 1932

Brillouin L (1900) Les réductions de la pesanteur au niveau de la mer; les différents géoides. Rev Gen Sci Pure Appl XI: 875 – 882

Bruns H (1876) Über einen Satz der Potentialtheorie. Crelle's J 81

Bruns H (1878) Die Figur der Erde. Königl Preuss Geodes Inst, Potsdam

Buchwaldt A (1916) Geometrisk bidrag til geoidens bestemmelse. Copenhagen

Buchwaldt A (1922) Les principes de la géodésie statique. Den Dan Graadmaaling, Kopenhagen

Burali-Forti C (1909) Alcune nuove espressioni assolute della curvatura in un punto di una superficie. R Accad Lincei 18: 1st sem, 50 – 55

Burali-Forti C (1912) Fondamenti della geometria differenziale su di una superficie col metodo vettoriale generale. Rend Circ Mat Palermo T XXXIII: 1st sem, 1 – 40

Burgatti P (1936) Sulle discontinuità degli scalari, vettori ed omografie vettoriali. Atti Accad Sci Bologna 40

Caldonazzo B (1924) Sulla geometria differenziale di una superficie. Rend Accad Lincei 33: Ser. 5, 396 – 400

Caldonazzo B (1942) Considerazioni geometriche sul potenziale gravitazionale. Rend Ist Lomb 75

Cassinis G (1927) Sulla determinazione dello schiacciamento terrestre mediante valori della gravità. Atti R Accad Torino 62

Cassinis G (1930) Sur l'adoption d'une formule internationale pour la pesanteur normale. Bull Géodés 26

Cassinis G (1932) Tables des valeurs de la pesanteur normale internationale. Bull Géodés 1932: no 36, 1941: no 32

Cassinis G (1934) Questioni di gravimetria. Pubbl Ist Topogr Geodes R Polit Milano

Christoffel EB (1865) Über die Bestimmung der Gestalt einer krummen Oberfläche durch lokale Messungen auf derselben. J Reine Angew Math T LXIV

Christoffel EB (1867) Über einige allgemeine Eigenschaften der Minimumsflächen. J Reine Angew Math Tome LXVII

Christoffel EB (1869) Allgemeine Theorie der geodätischen Dreiecke. Math Abh Akad Wiss, Berlin

Cisotti U (1941 a) Sistemi continui conservativi. Rend Accad Ital 2: Ser 8, 294 – 300

Cisotti U (1941 b) Campi tensoriali potenziali. Rend Accad Ital 2: Ser 8, 129 – 141

Cisotti U (1941 c) Invarianti quadratici dei tensori. Rend Accad Ital 2: Ser 8, 512 – 516

Darboux G (1889) Leçons sur la théorie générale des surfaces, vol II. pp 239 – 243 Gauthier-Villars, Paris

Dore P (1923 a) Sulle relazioni intercedenti fra le deviazioni del geoide da una superficie di riferimento poco diversa alla sfera e le anomalie della gravità. Mem Soc Astron Ital 2: no 3

Dore P (1923 b) Sul teorema di Clairaut. Boll Un Mat Ital, II: no 3

Dore P (1940 – 1941) Sull'uso della seconda funzioni di Green nei procedimenti di calcolo degli scostamenti del geoide dall'ellissoide a mezzo delle anomalie gravimetriche. Atti R Accad Sci Bologna, Ser IX Vol 8

Eötvös R (1908) Bericht über die geodätischen Arbeiten in Ungarn; Bestimmung der Gradienten der Schwerkraft und ihrer Niveauflächen mit Hilfe der Drehwaage. C R XV Conf Gén Assoc Géodés Int Berlin

Eötvös R (1909) Über geodätische Arbeiten in Ungarn, besonders über Beobachtungen mit der Drehwaage. C R XVI Conf Gen Assoc Geodes Int. London-Cambridge

Evjen HM (1936) The place of the vertical gradient in gravitational interpretation. Geophysics 1

Gorchkov GP (1934) On isogravitational surfaces. Bull Géodés no 43

Graaf Hunter J de (1935) The figure of the Earth from the gravity observations and the precision obtainable. Philos Trans R Soc London Ser A, 234, no 743

Gulatee BL (1933) The figure of the Earth. Gerlands Beitr Geophys 38:426

Gulatee BL (1935) The boundary problems of potential theory and geodesy. Gerlands Beitr Geophys 46, no 1 – 2

Gutenberg B (1930) Schwere und Druck im Erdinnern. Gerlands Beitr Geophys T 26
Haalck H (1941) Die Frage der Messung des vertikalen Schweregradienten. Beitr Angew Geophys 9: H 2
Hammer S (1938) Investigation of the vertical gradient of gravity. Trans Am Un Natl Res Counc 1:74
Haurwitz B (1928) Über die Änderung der Schwere im Erdinnern. Gerlands Beitr Geophys 19
Haurwitz B (1931) Über die Änderung der Schwere im Erdinnern. Gerlands Beitr Geophys 28, no 1 – 3
Helmert F (1900) Zur Bestimmung kleiner Flächenstücke des Geoids aus Lotabweichungen mit Rücksicht auf Lotkrümmung. Sitzungsber K Akad Wiss Berlin 42
Helmert F (1902) Über die Reduktion der auf der physischen Erdoberfläche beobachteten Schwerebeschleunigungen auf ein gemeinsames Niveau. Sitzungsber K Preuss Akad Wiss, Berlin
Hirvonen RA (1934) The continental undulations of the geoid. Ist Geodes Finn, Helsinki
Idelson N, Malkin B (1931) Die Stokessche Formel in der Geodäsie als Lösung einer Randwertaufgabe. Gerlands Beitr Geophys 29
Jeffreys H (1931) An application of the free air reduction of gravity. Gerlands Beitr Geophys 31:38
Jeffreys H (1932) Figure of the Earth. Gerlands Beitr Geophys 36:206
Jeffreys H (1938) The determination of gravity anomalies from deflections of the vertical. Mon Not R Astron Soc Geophys Suppl 4
Jelstrup HS (1937) Un aperçu des forces gravitationnelles agissant sur la Terre, et la figure terrestre. Nor Mat Tidensskr 37:19, no 1
Joachimsthal F (1848) Sur les normales infiniment voisines d'une surface courbe. J Math Pure Appl XIII
Jolly PV (1878) Anwendung der gewöhnlichen Waage zur Bestimmung der Abnahme der Schwere mit der Höhe. Abh K Bayr Akad Wiss München 13
Jung K (1936) Bemerkung zur Potentialtheorie des Schwerekraftfeldes. Z Geophys 12
Lambert WD (1930) The reduction of observed values of gravity to sea level. Bull Géodés, 107 – 181, N 26
Lambert WD (1934) Note on the reduction of gravity observations to sea level for the purpose of determining the geoid. Bull Géodés no 41
Lambert WD (1937) The analogue of Stokes' formula for the Prey and Bouguer gravity anomaly. Gerlands Beitr Geophys 49
Lambert WD, Darling FW (1931) Tables for the theoretical gravity according to the new international formula. Bull Géodés 32
Laura E (1923) Sul calcolo dei limiti superiori dei resti nelle formule che danno la gravità alla superficie di un pianeta. Mem Soc Astron Ital 2 Nuovo Ser
Laura E (1925) Sopra il teorema di Stokes. Mem Soc Astron Ital 3
Levi-Civita T (1934) Terne di congruenze sopra una superficie ed estensione della trigonometria. Composit Math 1: fasc I
Mader K (1948) Die Bestimmung einer Geoiderhebung aus Messungen mit der Drehwaage nach Eötvös. Riv Geofis Pure Appl 13
Malkin N (1939) Sur la détermination du géoide d'après les observations gravimétriques. Astron J Sov Un 16
Manacorda T (1948a) Sulla discontinuità delle derivate di potenziale di semplice strato. Atti Accad Naz Lincei 5: fasc 1 and 2, July – August 1948
Manacorda T (1948b) Sulle discontinuità delle derivate del potenziale di semplice strato. Atti Accad Naz Lincei fasc 3 – 4, September – October 1948
Marussi A (1944a) Sulla livellazione trigonometrica geoidica. Universo, anno 25: no 5
Marussi A (1944b) Saggio del geoide nel Mediterraneo centrale e nella zona alpina. Boll Geodes no 5
Marussi A (1947) Sulla discontinuità delle derivate seconde del potenziale di volume. Boll Geodes no 3
Marussi A (1949) Fondements de géométrie différentielle absolue du champ potentiel terrestre. Bull Géodés December 1949: no 14
Marussi A (1950a) Sulla variazione con l'altezza dei raggi di curvatura principali nella teoria di Somigliana. Boll Geodes Sci Affini, anno VIII: no 1
Marussi A (1950b) Principi di Geodesia intrinseca applicati al campo di Somigliana. Boll Geodes Sci Affini, Vol IX: no 2 (Chap. III.1, this Vol.)
Marussi A (1950c) Sviluppi di Legendre generalizzati per una curva qualunque tracciata su di una superficie pure qualunque. Rend Accad Naz Lincei 8: Ser 8, fasc 4 (Chap. I.3, this Vol.)

Marussi A (1950d) Sviluppi di Legendre generalizzati nello spazio. Rend Accad Naz Lincei 8: Ser 8, fasc 5 (Chap. I.4, this Vol.)
Marussi A (1950e) Sviluppi di Legendre generalizzati per una curva qualunque dello spazio. Rend Accad Naz Lincei 9: Ser 8, fasc 1 – 2 (Chap. I.5, this Vol.)
Marussi A (1950f) Il primo problema fondamentale della Geodesia ampliato nel campo di Somigliana. Geofis Pura Appl 18 (dedicated to Prof Somigliana) (Chap. III.2, this Vol.)
Marussi A (1951a) Sul calcolo dei simboli di Christoffel per le proiezioni di Mercatore e di Gauss. Atti Ist Veneto Sci Lett Art CIX
Marussi A (1951b) Geodesia spaziale. Boll Geodes Sci Affini no 1
Marussi A (1951c) Su alcune proprietà fondamentali delle rappresentazioni conformi fra superfici. Boll Geodes Sci Affini no 2 (Chap. V.1, this Vol.)
Marussi A (1951d) Les principes de la géodésie intrinsèque. Bull Géodés no 19
Messerschmitt JB (1908) Die Schwerebestimmung an der Erdoberfläche. Vieweg und Sohn, Braunschweig
Michailov A (1936) Über die Anwendung der Formel von Stokes und die dabei gebrauchende Reduktion der Schwere. C R Comm Geodes Balt, 8th Sess, Tallinn
Mineo C (1910a) Sistemi geografici sulle superfice. Period Mat XXVI
Mineo C (1910b) Sulle superficie riferite ad un sistema geografico e sulla determinazione intrinseca del geoide. G Battaglini 48
Mineo C (1911) Sulle formule fondamentali per il confronto della superficie geoidica con l'ellissoide besseliano. G Battaglini 44
Mineo C (1913) Sur la détermination indirecte de la différence des longitudes astronomiques au moyen d'une latitude et de deux azimuts réciproques observés. Bull Astron (Paris) XXX
Mineo C (1914) Sui problemi della trigonometria sferoidica. Atti Accad Sci Nat Catania 7
Mineo C (1921) Sulla determinazione intrinseca del geoide. Mem Soc Astron Ital 2
Mineo C (1924) Sulla rappresentazioni geografiche di una superficie su di un'altra. Not Mem Mat 2: fasc 3
Mineo C (1930) Ricerche intorno al campo di gravità terrestre. Atti R Accad Sci Lett Belle Art Palermo 15: fasc 1
Mineo C (1931) Sulle nuove determinazioni di gravità relativa in Sicilia e su qualche considerazione generale. Atti R Accad Sci Belle Art Palermo 17
Mineo C (1935 – 36) Sulla geometria di una superficie poco differente da un ellissoide con applicazioni al caso della terra. Ann Mat Pura Appl Ser 4, T XIV, no 3 – 4
Mineo C (1937a) Sulla riduzione dei valori osservati della gravità al livello del mare. G Sci Mat Econ Palermo 39
Mineo C (1937b) Il potenziale terrestre e la riduzione dei valori osservati della gravità ad una superficie di riferimento. Ann Mat Pura Appl Ser IV, XVI
Mineo C (1940) Sul passaggio da uno ad un altro degli ellissoidi locali relativi ad una data regione del geoide, e sulle conseguenti variazioni delle coordinate ellissoidiche dei vertici della rete geodetica. R Accad Ital 2: Ser 7, fasc 6
Mineo C (1940 – 41) Sul trasporto di una rete di triangoli geodetici da uno ad un altro degli ellissoidi locali di una data regione del geoide. G Sci Mat Econ Palermo 42: no 1
Mineo C (1940) Sulla possibilità di sostituire un unico ellissoide a due o piu ellissoidi di riferimento, e sulla determinazione del geoide. Osserv Astron Palermo 9: Nuovo Ser, no 3
Numerov B (1929) Interrelation between local gravity anomalies and the derivatives of the potential. Z Geophys 5:58
Ölander VR (1931) Über die Beziehung zwischen Lotabweichung und Schwereanomalien sowie über das Lotabweichungssystem in Südfinnland. Finn Geod Inst 17, Helsinki
Olczak T (1948) The measurements with the Eötvös balance and the problem of determining the higher normal derivatives of the external gravity potential. Inst Geol Pol, Warsaw
Paci P Sopra le derivate terze della funzione potenziale di una superficie. Rend Circ Mat Palermo
Pizzetti P (1887) Contribuzione allo studio geometrico della superficie terrestre. G Soc Lett Convers Sci Genova 17: August – September 1887
Pizzetti P (1895a) Sur la réduction des latitudes et des longitudes astronomiques au niveau de la mer. Astron Nachr 138
Pizzetti P (1895b) Intorno alla effettiva determinazione della superficie di livello terrestre entro regione limitate. Rend Accad Lincei 4

Pizzetti P (1901) Un principio fondamentale nello studio delle superfici di livello terrestri. Atti Accad Lincei 5: II sem
Pizzetti P (1909) Sul significato geometrico del secondo parametro differenziale sopra una superficie qualunque. Rend Accad Lincei 18: Ser 5
Pizzetti P (1910 – 11) Sopra il calcolo teorico delle deviazioni del geoide dall'ellissoide. Atti Accad Torino 44
Pizzetti P (1917) Geometria delle superficie e geometria della sfera. Rend Accad Lincei 26: Ser 5
Poincaré H (1901) Les mesures de gravité et la géodésie. Bull Astron XVII
Poincaré H (1901 b) Sur les déviations de la verticale en géodésie. Bull Astron XVIII
Poynting JH (1878) On a method of using the balance with great delicacy and on its employment to determine the mean density of the Earth. Proc R Soc London Ser A28:2 – 35
Poynting JH (1892) On a determination of the mean density of the Earth and the gravitation constant by means of the common balance. Philos Trans A 182, 565 – 655
Prey A (1904) Über die Reduktion der Schwerebeobachtungen auf das Meeresniveau. Österr Akad Sitzungsber Kl II
Reina V (1889) Di alcune proprietà delle linee caratteristiche. Rend Accad Lincei 5
Reina V (1890 a) Sulla linee coniugate di una superficie. Rend Accad Lincei 6
Reina V (1890 b) Di alcune formule relative alla teoria delle superfici. Rend Accad Lincei 6
Reina V (1893) Sulla determinazione dei raggi di curvatura di una superficie per mezzo di misure locali sopra di essa. Rend Accad Lincei 2
Reina V (1917) Sulla determinazione di porzioni della superficie del geoide per mezzo di misure sopra di essa. Rend Accad Lincei 26
Scheel K, Dieselhorst H (1895) Bestimmung der Änderung der Schwere mit der Höhe. Phys Reichsanst Wiss Abh 2:185
Schwinner R (1934) Außenraum und Innenraum (Schlichtung des Streits um die Schwerkraftsreduktion). Z Geophys 10:240
Serini R (1911) Intorno ad alcune formule relative allo Studio geometrico differenziale della superficie matematica terrestre e alla sua effettiva costruzione dai risultati di osservazioni e misure. Atti Mem Accad Virg Mantova 3
Silva G (1928) Sulla definizione della gravità normale. Rend R Accad Lincei 3: Ser 6, December 1928 (also Publ no 9 R Osserv Astron Padova)
Silva G (1930) Sulle formule della gravità normale. Rend R Accad Lincei April 1930: Ser 6 (also Publ no 10 R Osserv Astron Padova)
Soler E (1929) Su alcune formule ordinariamente adoperate per la riduzione della gravità osservata al geoide sulla loro verifica sperimentall. Bull Géodés no 24
Soler E (1930) Due campagne gravimetriche sul Carso compiute dall'Istituto di Geodesia della R. Università di Padova. R Comm Geodes Ital
Soler E (1931 – 32) Sur certains problèmes de gravimétrie. Bull Géodés no 29 and 33
Somigliana C (1915) Sulle derivate seconde della funzione potenziale di superficie. Atti Accad Sci Torino 51
Somigliana C (1927 a) Sulla estensione del teorema di Clairaut. Atti R Ist Veneto 86
Somigliana C (1927 b) Sulle relazioni che esistono fra le costanti geoidiche ed i valori della gravità. Rend R Accad Lincei June 1927
Somigliana C (1927 c) Sulla determinazione delle costanti geoidiche mediante sole misure di gravità. Rend R Accad Lincei March 1927
Somigliana C (1928) Quelques nouveaux principes théoriques en gravimétrie. Bull Géodés no 18
Somigliana C (1929) Teoria generale del campo gravitazionale dell'ellissoide di rotazione. Mem Soc Astron Ital 4
Somigliana C (1931) Sulle linee di forza dei campi newtoniani simmetrici intorno ad un asse. Atti R Accad Sci Torino 56
Somigliana C (1933) Les expressions finies de la pesanteur normale. Bull Géodés no 38
Somigliana C (1934) Le relazioni lineari che esistono fra i valori della gravità sul geoide ellissoidico. Atti R Accad Sci Torino 69
Somigliana (1936) Memorie Scelte. Lattes, Torino
Sterneck R von (1882) Untersuchungen über die Schwere im Innern der Erde. Mitt K K Milit Geogr Inst, Wien
Sterneck R von (1883) Wiederholung der Untersuchungen über die Schwere im Innern der Erde. Mitt K K Milit Geogr Inst, Wien

Stokes GC (1849) On the variation of gravity at the surface of the Earth. Trans Cambridge Philos Soc 8 (also Math Phys Pap 2:104, 1883)

Thiesen MF (1890) Détermination de la variation de la pesanteur avec la hauteur. Trav Mem Bur Int Poids Mesures VII

Tonolo A (1939a) Sviluppi di Puiseux-Weingarten generalizatti. Rend R Accad Naz Lincei 29: Ser 6, 1st sem, fasc 10

Tonolo A (1939b) Trasporto delle coordinate geografiche e dell'azimut lungo un arco di linea qualunque di un ellissoide di rotazione. Rend R Accad Naz Lincei 29: Ser 6, 1st sem, fasc 11

Tortorici P (1928) Estensione di alcune formule di Weingarten ed applicazione al problema della forma della Terra. Atti R Accad Sci Lett Belle Art Palermo 15

Tortorici P (1936) Sugli sferoidi isogravitazionali. Pubbl Ist Topogr Geodes Palermo 1: no 2

Viterbi A (1908) Sulla determinazione degli elementi intrinseci fondamentali della superficie terrestre mediante misure locali. Rend Circ Mat Palermo T XXV

Viterbi A (1909 – 10) Alcune formule relative alle traiettorie ortogonali di una famiglia di superfici e applicazioni di esse allo studio della superficie di livello terrestre. Atti Ist Veneto 69

Wegener K (1939) Bemerkungen zu dem vertikalen Gradienten der Schwere. Z Geophys H 5 – 6

3 Generalized Legendre Expansions for Any Curve Whatever, Traced Out Upon Any Surface Whatever[1]

1. The Legendre expansion, or also Delambre's formula, are the names given in Geodesy to those series of the power of the arc s of a geodesic traced on the ellipsoid of rotation, which permit us to determine the geographic coordinates φ_1 and λ_1 (latitude and longitude) of the extremity P_1 of the geodesic of finite length s, and of azimuth α_1 at P_1, if we know the geographic coordinates φ_0 and λ_0 of the other extremity P_0, and the azimuth of the geodesic also at P_0.

Such expansions have been generalized in various forms by several authors, who have sought to free them from the restriction that the curve of transport must be a geodesic, or from the restriction that the surface must be that of an ellipsoid. Nevertheless, the most generalized form of the problem has not up till now been confronted, that which consists in studying the above Legendre expansions for any curve whatever, traced out on any surface whatever.

Hristov (1937) has considered in particular expansions in series of powers of the arc of a geodesic traced out on any surface whatever, but referred to an isothermal-isometric system of coordinates which can have no simple relationship with geographic coordinates as long as the surface is not one of rotation. In such expansions the problem of the transport of azimuth cannot be taken into account. Tonolo (1939a), on the other hand, has considered the case of any curve whatever traced out, however, on a rotational ellipsoid, and thus valid for the usual geographic coordinates. His expansions thus represent a first effective generalization of those of Legendre, and are not limited to geographic coordinates alone, but also allow for the transport of azimuth. The subject has subsequently been treated in two notes by Boaga (1942, 1941–42), who found expressions for the transport of curvilinear coordinates along the arc of a geodesic on any surface whatever, coordinates which can be identified with the geographic coordinates, and with isothermal-isometric coordinates on surfaces of rotation.

It has not seemed wholly without interest, however, to re-examine the problem, seeking a more general solution, and setting out from the start by considering the intrinsic geographic coordinates on any surface whatever, thus following the lead indicated by Pizzetti, Reina, Mineo (Mineo 1910), and by other geodesists. In this investigation, account is also taken of my recent studies (Marussi 1948), and the methods of the tensor calculus are employed, thus

[1] Originally published as: Marussi A (1950) Sviluppi di legendre generalizzati per una curva qualunque tracciata su di una superficie pure qualunque. Rend Accad Naz Lincei 8: Ser 8, fasc 4, 342–347

conveying a facility and transparency to the formulas, and allowing them to be written in a more synthetic way without in any sense limiting their generality. The expansions found will handle also the transport of the astronomic azimuth along the given curve, the azimuth being defined on any surface whatever by the use of the spherical image of the surface.

2. Let us consider any regular convex surface Σ, which we shall take as closed, and a direction fixed in space, which we shall represent by the versor ω. Let us draw on the surface itself a double family of coordinate lines which we shall call $\varphi=$ const and $\lambda=$ const, such that (a) along the lines $\varphi=$ const the versor $\boldsymbol{n}$ normal to the surface, and directed towards its convexity, forms a constant angle $90-\varphi$ with the versor ω, (b) along the lines $\lambda=$ const the versor $\boldsymbol{n}$ is now held parallel to a fixed plane containing ω, a plane which forms in turn a dihedral angle of magnitude λ (reckoned positive in an anti-clockwise sense by an observer looking along the direction of ω) with a fixed plane (the fundamental meridian) passing through ω but otherwise arbitrarily chosen.

Given this, we can say that we have referred the surface to an intrinsic system of geographic coordinates φ and λ, with respect to the versor ω (the Earth's axis), and to the selected fundamental meridian.

If, for ease of writing, we put $\varphi=y^1$, $\lambda=y^2$, and if we denote by g_{ij} the fundamental metric tensor of the surface, we shall have at every point P of Σ the first fundamental form

$$dP \times dP = ds^2 = g_{ij} dy^i dy^j , \tag{1}$$

where from now on it is implied that the indices can assume the values 1 and 2, and where it is also implied that summation is carried out for these values, with respect to those indices which appear twice in any expression, one above and one below (dummy indices).

From the definition itself of the geographic coordinates on the considered surface, we have immediately that the square of the linear element of its spherical image is given by the third fundamental form

$$dn \times dn = d\varphi^2 + \cos^2\varphi d\lambda^2 . \tag{2}$$

We shall also write down the second fundamental form of Σ

$$-dn \times dP = D d\varphi^2 + 2D' d\varphi d\lambda + D'' d\lambda^2 \tag{3}$$

and we shall recall that, from the particular choice of the coordinates which, as well as fixing the metric of the surface, at the same time determine the direction of the normal with respect to the surrounding space, it follows that the coefficients of the first and second fundamental forms are related by (Marussi 1948)

$$g_{11} = D^2 + \frac{D'^2}{\cos^2\varphi} ; \qquad g_{12} = g_{21} = D'\left(D + \frac{D''}{\cos^2\varphi}\right);$$

$$g_{22} = D'^2 + \frac{D''^2}{\cos^2\varphi} . \tag{4}$$

We must now consider at each point of the surface an orthogonal triad of versors $(\boldsymbol{i}_1, \boldsymbol{i}_2, \boldsymbol{i}_3)$ of which $\boldsymbol{i}_3$ coincides with $\boldsymbol{n}$, $\boldsymbol{i}_1$ and $\boldsymbol{i}_2$ are tangential (normal to $\boldsymbol{n}$), with the first contained in the plane formed by the versors $\boldsymbol{n}$ and ω, in such a way that it forms an acute angle with ω, and the second normal to the other two, so that the triad is right-handed. We shall also denote by azimuth the angle on the plane $(\boldsymbol{i}_1, \boldsymbol{i}_2)$ measured in the sense that carries $\boldsymbol{i}_1$ to $\boldsymbol{i}_2$ (clockwise direction), and reckoned from $\boldsymbol{i}_1$.

If we now denote by $\boldsymbol{v}_1$ and $\boldsymbol{v}_2$ the vectors derived from the point P with respect to the geographic coordinates, vectors which are obviously tangential

$$\boldsymbol{v}_1 = P_{/1} = \frac{\partial P}{\partial \varphi}; \qquad \boldsymbol{v}_2 = P_{/2} = \frac{\partial P}{\partial \lambda} \tag{5}$$

we have the relationships (Marussi 1948)

$$\left.\begin{aligned} \boldsymbol{v}_1 &= -D\boldsymbol{i}_1 - \frac{D'}{\cos\varphi}\boldsymbol{i}_2 \\ \boldsymbol{v}_2 &= -D'\boldsymbol{i}_1 - \frac{D''}{\cos\varphi}\boldsymbol{i}_2 \end{aligned}\right\} . \tag{6}$$

3. The conditions of integrability of the surface Σ, also called the equations of Mainardi-Codazzi, which the coefficients D, D', D'' of the second form must satisfy, are then written, as Mineo (1910) has shown

$$\left.\begin{aligned} &\frac{\partial D}{\partial \lambda} - \frac{\partial D'}{\partial \varphi} + D'\operatorname{tg}\varphi = 0 \\ &\frac{\partial D''}{\partial \varphi} - \frac{\partial D'}{\partial \lambda} + D\sin\varphi\cos\varphi + D''\operatorname{tg}\varphi = 0 \end{aligned}\right\} . \tag{7}$$

Along with the vectors $\boldsymbol{v}_1$ and $\boldsymbol{v}_2$ it is convenient to consider also the reciprocal vectors $\boldsymbol{v}^1$ and $\boldsymbol{v}^2$, such that

$$\boldsymbol{v}^i \times \boldsymbol{v}_j = \delta^i_j \tag{8}$$

where δ^i_j is Kronecker's symbol, equal to unity for $i = j$, and equal to zero for $i \neq j$. One has

$$\left.\begin{aligned} \boldsymbol{v}^1 &= -\frac{D''}{\Delta}\boldsymbol{i}_1 + \frac{D'\cos\varphi}{\Delta}\boldsymbol{i}_2 \\ \boldsymbol{v}^2 &= \frac{D'}{\Delta}\boldsymbol{i}_1 - \frac{D\cos\varphi}{\Delta}\boldsymbol{i}_2 \end{aligned}\right\} \tag{9}$$

with $\Delta = DD'' - D'^2$.

Given any surface vector $\boldsymbol{u}$, we shall also write

$$\left.\begin{aligned} \boldsymbol{u} \times \boldsymbol{v}_k &= u_k \qquad & \boldsymbol{u} \times \boldsymbol{v}^k &= u^k \\ \boldsymbol{u} &= u_k \boldsymbol{v}^k \qquad & \boldsymbol{u} &= u^k \boldsymbol{v}_k \end{aligned}\right\} \tag{10}$$

and we shall call the u_k the covariant components of $\boldsymbol{u}$, and the u^k its contravariant components.

The Christoffel symbols of the second kind $\Gamma^i_{jk} = \Gamma^i_{kj} = \dfrac{\partial^2 P}{\partial y^j \partial y^k} \times v^i$ are then written (see Mineo 1910, Marussi 1948)

$$\left.\begin{aligned}
\Gamma^1_{11} &= \frac{1}{\Delta}\left(D''\frac{\partial D}{\partial\varphi} - 2D'\frac{\partial D'}{\partial\lambda} + D'\frac{\partial D}{\partial\lambda}\right);\\
\Gamma^1_{12} = \Gamma^1_{21} &= \frac{1}{\Delta}\left(D''\frac{\partial D}{\partial\lambda} - D'\frac{\partial D''}{\partial\varphi}\right);\\
\Gamma^1_{22} &= \frac{1}{\Delta}\left(-D''\frac{\partial D''}{\partial\varphi} + 2D''\frac{\partial D'}{\partial\lambda} - D'\frac{\partial D''}{\partial\lambda}\right) - \sin\varphi\cos\varphi;\\
\Gamma^2_{11} &= \frac{1}{\Delta}\left(-D'\frac{\partial D}{\partial\varphi} + 2D\frac{\partial D'}{\partial\lambda} - D\frac{\partial D}{\partial\lambda}\right);\\
\Gamma^2_{22} &= \frac{1}{\Delta}\left(D'\frac{\partial D''}{\partial\varphi} - 2D'\frac{\partial D'}{\partial\lambda} + D\frac{\partial D''}{\partial\lambda}\right);\\
\Gamma^2_{12} = \Gamma^2_{21} &= \frac{1}{\Delta}\left(D\frac{\partial D''}{\partial\varphi} - D'\frac{\partial D}{\partial\lambda}\right) + \mathrm{tg}\,\varphi
\end{aligned}\right\} . \tag{11}$$

Given all this, we propose to resolve the problem of determining the value of the geographic coordinates $y^i_{(1)}$ at a point P_1 of the surface Σ, knowing the coordinates $y^i_{(0)}$ of a pre-selected point P_0, and knowing a curve σ of finite length s traced on Σ, one end of which coincides with P_0, and the other with P_1; and also to determine the astronomical azimuth α_1 of the curve at P_1, given its azimuth α_0 at P_0. The curve will be defined, as well as by its length s, by its tangent $\boldsymbol{t}_0$ at P_0, and by its geodetic or tangential curvature γ which we shall take as given as a function of the arc.

If we now denote by $\boldsymbol{t}$ the versor tangent to the curve σ at the general point P, by $\boldsymbol{b}$ the versor on Σ normal to it, in such a way that the triad $(\boldsymbol{t}, \boldsymbol{b}, \boldsymbol{n})$ connected to the curve and to the surface is right-handed, then the generalized formulae of Frenet will hold

$$\left.\begin{aligned}
\frac{d\boldsymbol{t}}{ds} &= -\frac{1}{\rho_n}\boldsymbol{n} + \frac{1}{\rho_g}\boldsymbol{b}\\
\frac{d\boldsymbol{b}}{ds} &= -\frac{1}{\tau_g}\boldsymbol{n} - \frac{1}{\rho_g}\boldsymbol{t}
\end{aligned}\right\} \tag{12}$$

in which $1/\rho_n$ represents the normal curvature, $1/\rho_g = \gamma$ the geodetic or tangential curvature, and $1/\tau_g$ the geodetic torsion of σ at P.

Let us now denote by λ^i and ν^i respectively the contravariant components of $\boldsymbol{t}$ and of $\boldsymbol{b}$:

$$\left.\begin{aligned}
\lambda^i &= \boldsymbol{t}\times\boldsymbol{v}^i = \frac{dy^i}{ds}; \qquad & \nu^i &= \boldsymbol{b}\times\boldsymbol{v}^i;\\
\boldsymbol{t} &= \lambda^i\boldsymbol{v}_i = \lambda^i\frac{\partial P}{\partial y^i}; \qquad & \boldsymbol{b} &= \nu^i\boldsymbol{v}_i = \nu^i\frac{\partial P}{\partial y^i};
\end{aligned}\right\} \tag{13}$$

differentiating these with respect to s, and multiplying at the same time in a scalar manner by $\boldsymbol{v}^i$, taking account of the above formulae of Frenet, of Eqs. (8) and (13) and of the definition given for Christoffel's symbols of the second kind, we have immediately

$$\left.\begin{aligned} \frac{d\lambda^i}{ds} &= \gamma v^i - \Gamma^i_{jh}\lambda^j\lambda^h \\ \frac{dv^i}{ds} &= -\gamma\lambda^i - \Gamma^i_{jh}v^j\lambda^h \end{aligned}\right\} . \tag{14}$$

These are the differential equations of the considered curve σ with respect to the chosen coordinates; such equations we find also in Tonolo (1939b).

4. After this, the problem posed can be readily solved by means of series expansions quite analogous to those of Legendre for the ellipsoid, writing

$$y^i = y^i_{(0)} + s\lambda^i_{(0)} + \frac{s^2}{2!}\left(\frac{d\lambda^i}{ds}\right)_0 + \frac{s^3}{3!}\left(\frac{d^2\lambda^i}{ds^2}\right)_0 + \frac{s^4}{4!}\left(\frac{d^3\lambda^i}{ds^3}\right)_0 + \dots \tag{15}$$

where it only remains to determine the expressions for the successive derivatives of λ^i with respect to s, calculating their values at the point P_0.

We have, by a somewhat complicated calculation, but one of no real difficulty, and taking account of Eq. (14)

$$\left.\begin{aligned} \frac{d^2\lambda^i}{ds^2} &= \frac{d\gamma}{ds}v^i - \gamma^2\lambda^i - 3\gamma\Gamma^i_{jh}v^j\lambda^h + \left(2\Gamma^i_{hr}\Gamma^r_{jk} - \frac{\partial\Gamma^i_{jh}}{\partial y^k}\right)\lambda^j\lambda^h\lambda^k; \\ \frac{d^3\lambda^i}{ds^3} &= \frac{d^2\gamma}{ds^2}v^i - \frac{d\gamma}{ds}(3\gamma\lambda^i + 4\Gamma^i_{jh}v^j\lambda^h) - \gamma^3 v^i + \gamma^2\Gamma^i_{jh}(4\lambda^j\lambda^h \\ &\quad - 3v^j v^h) + \gamma\left\{\Gamma^i_{hr}\Gamma^r_{jk}(7v^j\lambda^k + 5v^h\lambda^j) - \frac{\partial\Gamma^i_{jk}}{\partial y^h}(5v^j\lambda^h \right. \\ &\quad \left. + \lambda^j v^h)\right\}\lambda^k + \left(2\Gamma^i_{lr}\frac{\partial\Gamma^r_{jh}}{\partial y^k} + 4\Gamma^r_{jh}\frac{\partial\Gamma^i_{rl}}{\partial y^k} + \Gamma^r_{jh}\frac{\partial\Gamma^i_{kl}}{\partial y^r} \right. \\ &\quad \left. - \frac{\partial^2\Gamma^i_{jh}}{\partial y^k\partial y^l} - 4\Gamma^i_{lr}\Gamma^r_{ks}\Gamma^s_{jh} - 2\Gamma^i_{rs}\Gamma^r_{hk}\Gamma^s_{jl}\right)\lambda^j\lambda^h\lambda^k\lambda^l \end{aligned}\right\} . \tag{16}$$

Calculation of the successive derivatives becomes naturally much more complex, but the terms written above will already permit a very high precision. In the case of the terrestrial ellipsoid, termination of the series with the terms written will at least guarantee, for geodesics no longer than 100 km, an accuracy of one-thousandth of a second of arc; and one hundreth of a second of arc for lengths not exceeding 150 km. When the given curve is a geodesic, all terms in Eqs. (14) and (16) which contain the factor γ or its derivatives will obviously vanish. It only remains to observe that the Eqs. (13), (14) and (16) are valid for whatever curvilinear coordinates may be chosen on Σ.

If the curve σ is defined at P_0 by its astronomic azimuth α_0, we have

$$\left.\begin{aligned} \boldsymbol{t}_0 &= \boldsymbol{i}_1 \cos\alpha_0 + \boldsymbol{i}_2 \sin\alpha_0 \\ \boldsymbol{b}_0 &= -\boldsymbol{i}_1 \sin\alpha_0 + \boldsymbol{i}_2 \cos\alpha_0 \end{aligned}\right\} \tag{17}$$

and thus

$$\left.\begin{aligned} \lambda^1_{(0)} &= \boldsymbol{t}_0 \times \boldsymbol{v}^1_{(0)} = -\frac{D''\cos\alpha_0}{\Delta} + \frac{D'\cos\varphi_0 \sin\alpha_0}{\Delta} \\ \lambda^2_{(0)} &= \boldsymbol{t}_0 \times \boldsymbol{v}^2_{(0)} = \frac{D'\cos\alpha_0}{\Delta} - \frac{D\cos\varphi_0 \sin\alpha_0}{\Delta} \\ v^1_{(0)} &= \boldsymbol{b}_0 \times \boldsymbol{v}^1_{(0)} = \frac{D''\sin\alpha_0}{\Delta} - \frac{D'\cos\varphi_0 \cos\alpha_0}{\Delta} \\ v^2_{(0)} &= \boldsymbol{b}_0 \times \boldsymbol{v}^2_{(0)} = -\frac{D'\sin\alpha_0}{\Delta} - \frac{D\cos\varphi_0 \cos\alpha_0}{\Delta} \end{aligned}\right\} \tag{18}$$

where it is implicit that the values of D, D', D'' and Δ are calculated at the point P_0.

5. As for the transport of the astronomical azimuth α along σ, it is only necessary to consider that the azimuth itself is provided at a general point P by the relationships

$$\left.\begin{aligned} \cos\alpha &= \boldsymbol{t} \times \boldsymbol{i}_1 = \lambda^k \boldsymbol{v}_k \times \boldsymbol{i}_1 = -D\lambda^1 - D'\lambda^2 \\ \sin\alpha &= \boldsymbol{t} \times \boldsymbol{i}_2 = \lambda^k \boldsymbol{v}_k \times \boldsymbol{i}_2 = -\frac{D'\lambda^1 + D''\lambda^2}{\cos\varphi} \end{aligned}\right\} \tag{19}$$

and that the λ^k are transported along σ by means of the series expressions

$$\lambda^k_{(1)} = \lambda^k_{(0)} + s\left(\frac{d\lambda^k}{ds}\right)_0 + \frac{s^2}{2!}\left(\frac{d^2\lambda^k}{ds^2}\right)_0 + \frac{s^3}{3!}\left(\frac{d^3\lambda^k}{ds^3}\right) + \dots \tag{20}$$

whose coefficients are those already calculated. Having thus obtained the values of $\lambda^k_{(1)}$ at P_1, and knowing at the same point the values of D, D', D'', the calculation of the azimuth α_1, reciprocal to α_0, is immediate.

In this way the problem posed at the beginning is completely resolved.

Bibliography

Boaga G (1942) Sul trasporto delle coordinate curvilinee lungo un arco di geodetica su di una superficie qualunque, con applicazioni al problema geodetico del trasporto delle coordinate geografiche. Universo anno 23: no 7, September 1942

Boaga G (1941 – 42) Il trasporto delle coordinate curvilinee lungo un arco di geodetica in alcuni casi particolari interessanti la Geodesia. Rend R Ist Lomb Sci Lett 75: fasc II

Hristov WK (1937) Berechnung der Koordinatendifferenz und der Ordinatenkonvergenz aus der Länge und dem Richtungswinkel einer geodätischen Strecke für eine beliebige Fläche und ein beliebiges isothermes Koordinatensystem. Z Vermessungswes 66:H6

Marussi A (1948) Fondements de géométrie différentielle absolue du champ potentiel terrestre. Commun Present VIII Assembl Gen Assoc Int Géodésie, Oslo

Mineo C (1910) Sulle superfici riferite ad un sistema geografico e sulla determinazione intrinseca del geoide. G Battaglini 48

Tonolo A (1939a) Trasporto delle coordinate geografiche e dell'azimut lungo un arco di linea qualunque di un ellissoide di rotazione. Rend R Accad Naz Lincei 29: Ser 6a, 1st sem, fasc II, June 1939

Tonolo A (1939b) Sviluppi di Puiseux-Weingarten generalizzati. Rend R Accad Naz Lincei 29: Ser 6, 1st sem, fasc 10, March 1939

4 Generalized Legendre Expansions in Space[1]

1. In a preceding chapter, we recalled the Legendre expansions which yield series of powers of the arc s of a geodesic traced out on an ellipsoid of rotation, and which are able to provide the geographic coordinates φ and λ (latitude and longitude) and the azimuth α at one end of the geodesic, if we know the corresponding elements at the other end. It was noted that these expansions have been generalized in various ways by several authors, and there was presented a further generalization for any curve traced out on any surface whatever.

The consideration of any surface whatever is of notable importance in Geodesy, insofar as this is concerned with the study of an equipotential surface of gravity (geoid), presumed to be known by the experimental determination of its differential elements. If, however, we take account of the fact that geodetic measurements take place only exceptionally on a single equipotential surface, (for this presumes the existence of extensive plains, while as a rule the station points are found at very different levels, and hence on different equipotential surfaces), we are faced with the problem of generalizing the above-mentioned fundamental expansions to the case which occurs as the rule in Geodesy.

In operational Geodesy, the problem is resolved by the so-called reduction of the linear and angular measurements to the geoid (or to the ellipsoid); but this operation takes place by introducing simplifying hypotheses which do not correspond to reality and which thereby falsify the results of the operations, thus taking away their logical rigour.

In this chapter it is proposed to show how these difficulties can be circumvented by considering the simultaneous transport of the geographic coordinates, of azimuth, of dynamic height, and of zenith distance, from one point to another of the gravity field, along finite curves of space geodesics, and thus along segments of straight lines, in such a way as to avoid the need to apply arbitrary reductions to observed values.

The relationships which will be established will also then find application both for the actual field that is available for observation, as well as for the normal field, established for example on the basis of Somigliana's theory, and adapted for specific applications by Silva and by Cassinis.

In this regard we should note that at the present time in Geodesy there is a move towards measuring, as well as the lengths of surface geodesics (geodetic bases, sides of triangulation), also the real distance between points of the Earth's

[1] Originally published as: Marussi A (1950) Sviluppi di legendre generalizzati nello spazio. Rend Accad Naz Lincei 8: Ser 8, fasc 5, 486 – 489

surface, along optical trajectories which to a high degree of approximation can be taken as segments of straight lines. The instruments conceived for this purpose, or which can be adapted to it, are the geodimeter of Bergstrand, based on an interference method using light waves, and the methods based on Radar, which use very short electromagnetic waves.

2. In a previous paper (Marussi 1949), the terrestrial gravity field has been referred to the coordinates most suited to its study from the geodetic viewpoint, by referring every point P on one of its equipotential surfaces to its astronomical latitude φ, and to its astronomical longitude λ; and by distinguishing one equipotential surface from another by means of the value W of the potential which characterizes it. As is well known, the potential difference between two surfaces measures, at least to an inessential constant factor, what is called in Geodesy the difference in their dynamic heights. The values of φ, λ and W, which for reasons of symmetry we shall also denote by y^1, y^2, y^3 in that order, then represent the coordinates of the point P.

In such a system of curvilinear coordinates, the fundamental covariant metric vectors $\boldsymbol{v}_r = \partial P/\partial y^r$ are given by

$$\left.\begin{aligned} \boldsymbol{v}_1 &= \frac{\partial P}{\partial y^1} = -D\boldsymbol{i}_1 - \frac{D'}{\cos\varphi}\boldsymbol{i}_2 \\ \boldsymbol{v}_2 &= \frac{\partial P}{\partial y^2} = -D'\boldsymbol{i}_1 - \frac{D''}{\cos\varphi}\boldsymbol{i}_2 \\ \boldsymbol{v}_3 &= \frac{\partial P}{\partial y^3} = -H_1\boldsymbol{i}_1 - H_2\boldsymbol{i}_2 - \frac{1}{g}\boldsymbol{i}_3 \end{aligned}\right\} \tag{1}$$

and the contravariant $\boldsymbol{v}^r = \operatorname{grad} y^r$ by

$$\left.\begin{aligned} \boldsymbol{v}^1 &= -\frac{D''}{\Delta}\boldsymbol{i}_1 + \frac{D'\cos\varphi}{\Delta}\boldsymbol{i}_2 + \frac{gh\cos\varphi}{\Delta}\boldsymbol{i}_3 \\ \boldsymbol{v}^2 &= \frac{D'}{\Delta}\boldsymbol{i}_1 - \frac{D\cos\varphi}{\Delta}\boldsymbol{i}_2 + \frac{gk\cos\varphi}{\Delta}\boldsymbol{i}_3 \\ \boldsymbol{v}^3 &= -g\boldsymbol{i}_3 \end{aligned}\right\} \tag{2}$$

in which D, D', D'' are as usual the coefficients of the second form of the equipotential surface through P, referred to the variables φ and λ, $\Delta = DD'' - D'$, g the modulus of gravity, $h = (D''H_1)/(\cos\varphi) - D'H_2$, and $k = DH_2 - (D'H_1)/(\cos\varphi)$, as for H_1 and H_2, are coefficients related to the gravimetric gradient; $\boldsymbol{i}_3$ is the versor of the external normal to the equipotential surface, $\boldsymbol{i}_1$ the tangential versor directed to the astronomic north, and $\boldsymbol{i}_2$ finally forming with $\boldsymbol{i}_1$ and $\boldsymbol{i}_3$ a unit orthogonal right-handed triad.

In Marussi (1949/1950) were given moreover the expressions for Christoffel's symbols of the second kind Γ^r_{ij}, which serve to connect one metric triad with another infinitesimally close.

3. Given this, let us consider two points P_0 and P in space, and the vector $P - P_0 = \boldsymbol{t}s$ which joins them, where $\boldsymbol{t}$ is a unit vector. At P_0 we shall have

$$t = \boldsymbol{i}_1 \sin z \cos\alpha + \boldsymbol{i}_2 \sin z \sin\alpha + \boldsymbol{i}_3 \cos z \tag{3}$$

where it is assumed that all scalar and vectorial elements refer to the point P_0, and α and z are then as usual the azimuth of the vertical plane through P, and the zenith distance.

We have immediately

$$t = \frac{dP}{ds} = v_i \frac{dy^i}{ds} = v_i \lambda^i; \qquad \lambda^i = t \times v^i = \frac{dy^i}{ds}; \tag{4}$$

from which it is seen that the λ^i are the contravariant components of t, and are given, writing out also the values of h and k, by

$$\left.\begin{aligned} \lambda^1 &= -\frac{D''}{\Delta}\sin z\cos\alpha + \frac{D'}{\Delta}\cos\varphi\sin z\sin\alpha \\ &\quad + \frac{1}{\Delta}\left(D'\frac{\partial \lg g}{\partial\lambda} - D''\frac{\partial \lg g}{\partial\varphi}\right)\cos z \\ \lambda^2 &= \frac{D'}{\Delta}\sin z\cos\alpha - \frac{D}{\Delta}\cos\varphi\sin z\sin\alpha \\ &\quad + \frac{1}{\Delta}\left(D'\frac{\partial \lg g}{\partial\varphi} - D\frac{\partial \lg g}{\partial\lambda}\right)\cos z \\ \lambda^3 &= -g\cos z \end{aligned}\right\}. \tag{5}$$

After this we shall have, expanding in a Taylor's series

$$y^r = y^r_{(0)} + s\left(\frac{dy^r}{ds}\right)_0 + \frac{s^2}{2!}\left(\frac{d^2y^r}{ds^2}\right)_0 + \frac{s^3}{3!}\left(\frac{d^3y^r}{ds^3}\right)_0 + \dots \tag{6}$$

in which the subscripts remind us that the values of the derivatives must be calculated at the point P_0.

To carry out such a calculation, we must have recourse to the equations of geodesics in general coordinates; as is well known, these equations are written, in the usual conventions of the tensor calculus,

$$\frac{d^2y^r}{ds^2} = \frac{d\lambda^r}{ds} = -\Gamma^r_{ij}\lambda^i\lambda^j \tag{7}$$

in which the summation is obviously taken for each index over the values 1 to 3.

Differentiating further we have, by a tedious but not difficult calculation,

$$\left.\begin{aligned} \frac{d^3y^r}{ds^3} &= \frac{d^2\lambda^r}{ds^2} = \left(2\Gamma^r_{hi}\Gamma^i_{jk} - \frac{\partial\Gamma^r_{jh}}{\partial y^k}\right)\lambda^j\lambda^h\lambda^k; \\ \frac{d^4y^r}{ds^4} &= \frac{d^3\lambda^r}{ds^3} = \left(2\Gamma^r_{li}\frac{\partial\Gamma^i_{jh}}{\partial y^k} + 4\Gamma^i_{jh}\frac{\partial\Gamma^r_{il}}{\partial y^k} + \Gamma^i_{jh}\frac{\partial\Gamma^r_{kl}}{\partial y^i}\right. \\ &\quad \left. - \frac{\partial^2\Gamma^r_{jh}}{\partial y^k\,\partial y^l} - 4\Gamma^r_{li}\Gamma^i_{ks}\Gamma^s_{jh} - 2\Gamma^r_{is}\Gamma^i_{hk}\Gamma^s_{jl}\right)\lambda^j\lambda^h\lambda^k\lambda^l \end{aligned}\right\}. \tag{8}$$

Finally, substituting the values, found for the λ^r and their observations at the point P_0, in the Taylor expansion, we obtain the desired formulae to solve the first part of the problem, that of the transport of the geographic coordinates and of the dynamic height.

4. As regards the transport of azimuth and zenith distance, we can establish analogous series expansions

$$\lambda^r = \lambda^r_{(0)} + s\left(\frac{d\lambda^r}{ds}\right)_0 + \frac{s^2}{2!}\left(\frac{d^2\lambda^r}{ds^2}\right)_0 + \frac{s^3}{3!}\left(\frac{d^3\lambda^r}{ds^3}\right) + \dots \tag{9}$$

in which the values of the derivatives are the same as those already used, and simply have to be multiplied by a different power of the arc s. Knowing the values of the λ^r at P by means of the formula written immediately above, the calculation of the reciprocal azimuth α and zenith distance z is straightforward; we have immediately, from (1), (3) and (4)

$$\left.\begin{aligned} \operatorname{tg}\alpha &= \frac{1}{\cos\varphi}\,\frac{D'\lambda^1 + D''\lambda^2 + \lambda^3(\partial/\partial\lambda)(1/g)}{D\lambda^1 + D'\lambda^2 + \lambda^3(\partial/\partial\varphi)(1/g)} \\ \cos z &= -\frac{1}{g}\lambda^3 \end{aligned}\right\} \tag{10}$$

noting that $H_1 = (\partial/\partial\varphi)(1/g)$ and $H_2 = (1/\cos\varphi)(\partial/\partial\lambda)(1/g)$, and where obviously all the values are taken as calculated at the point P.

It will be seen once more that in the developments given here there is no longer any need for a distinction between the azimuth of the geodesic and the azimuth of the normal section, a distinction which it is necessary to take into account in reducing the observed azimuths to those which are to be introduced into the usual calculation of the transport of geographic coordinates on a surface, in so much as here the geodesic is itself contained in the reciprocal normal sections which have been considered.

Bibliography

Marussi A (1949) Fondements de géométrie differentielle absolue du champ potentiel terrestre. Bull Géodés Assoc Int Géodésie, Paris, no 14, December 1949

Marussi A (1950) Sviluppi di Legendre generalizzati per una curva qualunque tracciata su di una superficie pure qualunque. Rend Accad Naz Lincei 8: fasc 4, 342–347 (Chap. I.3, this Vol.)

5 Generalized Legendre Expansions for Any Curve Whatever in Space[1]

1. The preceding chapter considered the transport of geographic coordinates and azimuth, of dynamic height and zenith distance between points separated by a finite distance, and belonging to two different equipotential surfaces of the Earth's gravity field, taking as the line of transport a space geodesic, i.e., a straight line. The expansions obtained represent a generalization of those known in Geodesy as Legendre expansions (or also as formulae of Delambre), and their interest lies in the fact that these expansions extended into space open up a new way of dealing with reductions to the geoid without introducing superfluous hypotheses. They show how it is possible to develop with complete generality the problem of trigonometric levelling, and to deal with the trigonometric transport of the dynamic height and zenith distance which are intimately linked to geographic coordinates and azimuth. The same expansions moreover offer the possibility of determining experimentally the values of the coefficients which appear in them, and which define the characteristics of the field under examination.

Specifying such a field, and taking, for example, the normal field of Somigliana as a reference, in the form in which it has been internationally adopted following the work and the proposals of Cassinis and of Silva, one can obtain conventional formulae which are particularly simple and apt for practical calculations.

The preceding chapter also stressed the modern tendency in Geodesy to measure large distances no longer along the geodesics of an equipotential surface, as was exclusively done in the past, but rather along optical trajectories (the geodimeter of Bergstrand), or those of very short radio waves (Radar) in space, trajectories which to a first approximation can be taken as straight lines, and thus as space geodesics.

As a further extension of these investigations, we have the problem of generalizing the curve of transport in space, assuming any sort of curved or twisted line. This can have immediate practical application, insofar as such a line can, for example be identified with the trajectories of microwaves or of optical rays when the refraction due to the presence of the atmosphere is taken into account.

2. In Chapter I.4 we have already provided tensor expressions for the Earth's gravity field referred to coordinates y^1, y^2, y^3, which are equal respectively to the geographic latitude φ, to the longitude λ, and to the difference in dynamic height

[1] Originally published as: Marussi A (1950) Sviluppi di legendre generalizzati per una curva qualunque dello spazio. Rend Accad Naz Lincei 9: Ser 8, fasc 1 – 2, 80 – 83

W measured in units of work. To these expressions we may add (Marussi 1949), the expressions found for the Christoffel symbols of the second kind relative to these same coordinates

$$\left.\begin{aligned}
&\Gamma^1_{11}=\frac{1}{\Delta}\left(D''\frac{\partial D}{\partial\varphi}-2D'\frac{\partial D'}{\partial\varphi}+D'\frac{\partial D}{\partial\lambda}+Dgh\cos\varphi\right);\\
&\Gamma^1_{12}=\Gamma^1_{21}=\frac{1}{\Delta}\left(D''\frac{\partial D}{\partial\lambda}-D'\frac{\partial D''}{\partial\varphi}+D'gh\cos\varphi\right);\\
&\Gamma^1_{22}=\frac{1}{\Delta}\left(2D''\frac{\partial D'}{\partial\lambda}-D''\frac{\partial D''}{\partial\varphi}-D'\frac{\partial D''}{\partial\lambda}+D''gh\cos\varphi\right)-\sin\varphi\cos\varphi;\\
&\Gamma^1_{13}=\Gamma^1_{31}=\frac{1}{\Delta}\left(D''\frac{\partial D}{\partial W}-D'\frac{\partial D'}{\partial W}\right);\\
&\Gamma^1_{23}=\Gamma^1_{32}=\frac{1}{\Delta}\left(D''\frac{\partial D'}{\partial W}-D'\frac{\partial D''}{\partial W}\right);\\
&\Gamma^1_{33}=\frac{1}{\Delta}\left(D''\frac{\partial H_1}{\partial W}-D'\frac{\partial H_2}{\partial W}\cos\varphi+\frac{\partial \lg g}{\partial W}h\cos\varphi\right);\\
&\Gamma^2_{11}=\frac{1}{\Delta}\left(2D\frac{\partial D'}{\partial\varphi}-D'\frac{\partial D}{\partial\varphi}-D\frac{\partial D}{\partial\lambda}+Dgk\cos\varphi\right);\\
&\Gamma^2_{12}=\Gamma^2_{21}=\frac{1}{\Delta}\left(D\frac{\partial D''}{\partial\varphi}-D'\frac{\partial D}{\partial\lambda}+D'gk\cos\varphi\right)+\operatorname{tg}\varphi;\\
&\Gamma^2_{22}=\frac{1}{\Delta}\left(D'\frac{\partial D''}{\partial\varphi}-2D'\frac{\partial D'}{\partial\lambda}+D\frac{\partial D''}{\partial\lambda}+D''gk\cos\varphi\right);\\
&\Gamma^2_{13}=\Gamma^2_{31}=\frac{1}{\Delta}\left(D\frac{\partial D'}{\partial W}-D'\frac{\partial D}{\partial W}\right);\\
&\Gamma^2_{23}=\Gamma^2_{32}=\frac{1}{\Delta}\left(D\frac{\partial D''}{\partial W}-D'\frac{\partial D'}{\partial W}\right);\\
&\Gamma^2_{33}=\frac{1}{\Delta}\left(D\frac{\partial H_2}{\partial W}\cos\varphi-D'\frac{\partial H_1}{\partial W}+\frac{\partial \lg g}{\partial W}k\cos\varphi\right);\\
&\Gamma^3_{11}=-gD;\quad \Gamma^3_{12}=\Gamma^3_{21}=-gD';\quad \Gamma^3_{22}=-gD'';\quad \Gamma^3_{33}=-\frac{\partial \lg g}{\partial W}\\
&\Gamma^3_{13}=\Gamma^3_{31}=\Gamma^3_{23}=\Gamma^3_{32}=0
\end{aligned}\right\}\quad(1)$$

where

$$\left.\begin{aligned} &\Delta = DD'' - D'^2; \quad H_1 = -\frac{1}{g}\,\frac{\partial \lg g}{\partial \varphi}; \quad H_2 = -\frac{1}{g\cos\varphi}\,\frac{\partial \lg g}{\partial \lambda};\\ &h = \frac{D''H_1}{\cos\varphi} - D'H_2; \quad k = DH_2 - \frac{D'H_1}{\cos\varphi} \end{aligned}\right\} \tag{2}$$

and g represents the modulus of gravity.

3. The curve σ of transport is presumed to be given by means of its following elements: the coordinates $y^r_{(0)}$ of one of its points P_0, the tangent $\boldsymbol{t}_{(0)}$ and the binormal $\boldsymbol{b}_{(0)}$ at this point, the curvature of flexion γ and the curvature of torsion τ expressed as a function of the arc s beginning from P_0. The problem consists in determining the coordinates y^r of any point P situated at a finite distance s along σ, and of the new positions taken up at P by the versors $\boldsymbol{t}$ (tangent) and $\boldsymbol{b}$ (binormal). As is seen, such a problem has been expanded by comparison with the foregoing results, for in addition to the transport of the geographic coordinates and of the height, azimuth, and zenith distance, we have now added that of the azimuth and zenith distance of the binormal, and for this reason the scalar elements to be transported now comprise six mutually independent quantities, exactly the degrees of freedom of a rigid body.

Denoting by $(\boldsymbol{t}, \boldsymbol{n}, \boldsymbol{b})$ the principal trihedron of σ, which we take as right-handed, and by λ^r, ν^r, β^r the contravariant components of their versors

$$\lambda^r = \boldsymbol{t} \times \boldsymbol{v}^r = \frac{dy^r}{ds}; \qquad \nu^r = \boldsymbol{n} \times \boldsymbol{v}^r; \qquad \beta^r = \boldsymbol{b} \times \boldsymbol{v}^r \tag{3}$$

in which the $\boldsymbol{v}^r$ represent the reciprocal vectors, as is explained in Chapter I.4, whose intrinsic derivatives are written, as the formulas of Frenet,

$$\frac{\delta\lambda^r}{\delta s} = \gamma\nu^r; \qquad \frac{\delta\beta^r}{\delta s} = \tau\nu^r; \qquad \frac{\delta\nu^r}{\delta s} = -\gamma\lambda^r - \tau\beta^r; \tag{4}$$

whence

$$\left.\begin{aligned} &\frac{\delta^2\lambda^r}{\delta s^2} = \frac{d\gamma}{ds}\nu^r - \gamma(\gamma\lambda^r + \tau\beta^r);\\ &\frac{\delta^3\lambda^r}{\delta s^3} = \frac{d^2\gamma}{ds^2}\nu^r - \frac{d\gamma}{ds}(3\gamma\lambda^r - 2\tau\beta^r) - \frac{d\tau}{ds}\gamma\beta^r - \gamma(\gamma^2 + \tau^2)\nu^r;\\ &\frac{\delta^2\beta^r}{\delta s^2} = \frac{d\tau}{ds}\nu^r - \tau(\gamma\lambda^r + \tau\beta^r) \end{aligned}\right\}. \tag{5}$$

Observing now that we have, for any simple contravariant tensor u^r

$$\frac{du^r}{ds} = \frac{\delta u^r}{\delta s} - u^i\lambda^j\Gamma^r_{ij}, \tag{6}$$

with a tedious but not difficult calculation we obtain

$$\left.\begin{aligned}
&\frac{d\lambda^r}{ds} = \gamma v^r - \lambda^i \lambda^j \Gamma^r_{ij}; \qquad \frac{d\beta^r}{ds} = \tau v^r - \beta^i \lambda^j \Gamma^r_{ij};\\
&\frac{d^2\lambda^r}{ds^2} = \frac{d\gamma}{ds} v^r - \gamma(\gamma\lambda^r + \tau\beta^r) - 3\gamma\Gamma^r_{ij}\lambda^i v^j + \left(2\Gamma^r_{is}\Gamma^s_{jh} - \frac{\partial\Gamma^r_{ij}}{\partial y^h}\right)\lambda^i\lambda^j\lambda^h;\\
&\frac{d^3\lambda^r}{ds^3} = \frac{d^2\gamma}{ds^2} v^r - \frac{d\gamma}{ds}(3\gamma\lambda^r + 2\tau\beta^r + 4\Gamma^r_{ij}v^i\lambda^j) - \gamma\frac{d\tau}{ds}\beta^r\\
&\qquad - \gamma(\gamma^2+\tau^2)v^r + \gamma^2\Gamma^r_{ij}(4\lambda^i\lambda^j - 3v^iv^j) + 4\gamma\tau\Gamma^r_{ij}\beta^i\lambda^j\\
&\qquad + \gamma\left(7\Gamma^r_{hs}\Gamma^s_{ij} + 5\Gamma^r_{is}\Gamma^s_{jh} - 5\frac{\partial\Gamma^r_{ij}}{\partial y^h} - \frac{\partial\Gamma^r_{hj}}{\partial y^i}\right) v^i\lambda^j\lambda^h\\
&\qquad - \left(2\Gamma^r_{ls}\Gamma^s_{jh}\Gamma^l_{ik} + 4\Gamma^r_{is}\Gamma^s_{lh}\Gamma^l_{jk} - 4\Gamma^s_{jh}\frac{\partial\Gamma^r_{is}}{\partial y^k} - 2\Gamma^r_{is}\frac{\partial\Gamma^s_{jh}}{\partial y^k}\right.\\
&\qquad \left. - \Gamma^l_{hk}\frac{\partial\Gamma^r_{ij}}{\partial y^l} + \frac{\partial^2\Gamma^r_{ij}}{\partial y^h\partial y^k}\right)\lambda^i\lambda^j\lambda^h\lambda^k;\\
&\frac{d^2\beta^r}{ds^2} = \frac{d\tau}{ds} v^r - \tau(\gamma\lambda^r + \tau\beta^r) - \Gamma^r_{ij}(2\tau v^i\lambda^j + \gamma\beta^i v^j)\\
&\qquad + \left(\Gamma^r_{is}\Gamma^s_{jh} + \Gamma^r_{js}\Gamma^s_{ih} - \frac{\partial\Gamma^r_{ij}}{\partial y^h}\right)\beta^i\lambda^j\lambda^h
\end{aligned}\right\} . \quad (7)$$

From this the usual taylor expansions provide the formulae sought

$$\left.\begin{aligned}
y^r &= y^r_{(0)} + s\lambda^r_{(0)} + \frac{s^2}{2!}\frac{d\lambda^r_{(0)}}{ds} + \frac{s^3}{3!}\frac{d^2\lambda^r_{(0)}}{ds^2} + \frac{s^4}{4!}\frac{d^3\lambda^r_{(0)}}{ds^3} + \dots\\
\lambda^r &= \lambda^r_{(0)} + s\frac{d\lambda^r_{(0)}}{ds} + \frac{s^2}{2!}\frac{d^2\lambda^r_{(0)}}{ds^2} + \frac{s^3}{3!}\frac{d^3\lambda^r_{(0)}}{ds^3} + \dots\\
\beta^r &= \beta^r_{(0)} + s\frac{d\beta^r_{(0)}}{ds} + \frac{s^2}{2!}\frac{d^2\beta^r_{(0)}}{ds^2} + \dots
\end{aligned}\right\} \quad (8)$$

in which for practical applications it should be sufficient to halt the series at the terms as written; and where it is understood that all the derivatives are to be calculated at P_0.

4. If the versors $\boldsymbol{t}$ and $\boldsymbol{b}$ are given, as will be the case in general, by means of the astronomic azimuths α_t and α_b of the normal sections which contain them, and by means of their zenith distances z_t and z_b, we shall have (see Chap. I.4)

$$\left.\begin{aligned}
\boldsymbol{t} &= \boldsymbol{i}_1 \sin z_t \cos\alpha_t + \boldsymbol{i}_2 \sin z_t \sin\alpha_t + \boldsymbol{i}_3 \cos z_t\\
\boldsymbol{b} &= \boldsymbol{i}_1 \sin z_b \cos\alpha_b + \boldsymbol{i}_2 \sin z_b \sin\alpha_b + \boldsymbol{i}_3 \cos z_b
\end{aligned}\right\} \quad (9)$$

from which are obtained expressions analogous to Eq. (5) of Chapter I.4, both for the λ^r and for the β^r, and these permit the calculation of the $\lambda^r_{(0)}$ and $\beta^r_{(0)}$ at

P_0. Conversely, and with procedures quite analogous to those described in depth in that chapter, it will be easy to calculate the values of α_t, z_t, α_b, z_b at P, once we know the values of λ^r and β^r at P itself. As for ν^r, it is sufficient to recall that $\nu^r = \boldsymbol{b} \wedge \boldsymbol{t} \times \boldsymbol{v}^r$.

The formulae found completely resolve the problem proposed. They become notably simplified if the curve of transport is plane, in which case we must everywhere set equal to zero the curvature of torsion τ and its derivatives. If we further take the curvature of flexion γ to be zero, they reduce finally to the formulae of Chapter I.4 for transport along a straight line.

Bibliography

Marussi A (1949) Fondements de géométrie différentielle absolue du champ potentiel terrestre. Bull Géodés Assoc Int Géodésique, Ncuv Ser, no 14, 411 – 439, December 1949

6 The Tidal Field of a Planet and the Related Intrinsic Reference Systems[1]

1. The Gravity Field

The gravitational accelerations acting on a body subject to the gravity field $\boldsymbol{g}$ of the Earth or of a planet and (1) suspended at its centre of gravity, or (2) in inertial motion around the planet, are the so-called tidal accelerations; they arise from the lack of uniformity of the gravity field. In case (1), $\boldsymbol{g}$ includes the centrifugal acceleration due to the rotation of the planet, and in case (2) the Newtonian gravitational acceleration only.

We shall place ourselves in the local point of view of the *Nahewirkungsgesetz* of Hermann Weyl describing the structure of the field at the observation point, disregarding the sources that produce the field. The ruling equations are then:

$$\operatorname{curl} \boldsymbol{g} = 0\,, \qquad \operatorname{div} \boldsymbol{g} = \begin{cases} 2\omega^2 - 4\pi G\delta & \text{(a)} \\ 0 & \text{(b)} \end{cases} \tag{1.1}$$

from which follows

$$\boldsymbol{g} = \operatorname{grad} W\,, \qquad \nabla^2 W = \begin{cases} 2\omega^2 - 4\pi G\delta & \text{(a)} \\ 0 & \text{(b)} \end{cases} \tag{1.1a}$$

according to whether (a) the observer is rotating with the Earth at the angular velocity ω, and there is a local distribution of matter of density δ (Poisson's equation); or else (b) the observer is *in vacuo* and not rotating (Laplace's equation). Furthermore, in order to describe the tensor quantities and geometrical objects that will be considered, we shall introduce a reference system intrinsically defined within the gravity (or gravitational) field of the planet itself. For these reasons the results at which we arrive will have a local-intrinsic character.

2. The Tidal Field

In cases (1) the suspended body and (2) the inertial motion, the resultant of the acting accelerations applied to the centre of gravity P_0 of a body is exactly balanced by (1) the tension of the suspension wire or by (2) the inertial

[1] Originally published as: Marussi A (1979) Geophys J R Astron Soc 56:409 – 417

accelerations. The only accelerations acting on any other point $P = P_0 + dP$ in the neighbourhood of P_0 are therefore given by

$$\frac{d\boldsymbol{g}}{ds} \cdot ds \qquad (ds = |dP|) .$$

Since

$$\boldsymbol{g} = -g\boldsymbol{n} , \quad g = |\boldsymbol{g}| \tag{2.1}$$

in which $\boldsymbol{n}$ is the unit vector of the vertical directed upwards, we have

$$\frac{d\boldsymbol{g}}{ds} = -g\left[\frac{d\boldsymbol{n}}{ds} + \boldsymbol{n}\left(\operatorname{grad}\ln g \cdot \frac{dP}{ds}\right)\right] . \tag{2.2}$$

By the generalized Bruns formula

$$\operatorname{grad}\ln g = \boldsymbol{f} - \left(H + \frac{2\omega^2 - 4\pi G\delta}{g}\right)\boldsymbol{n} = \boldsymbol{f} - H^*\boldsymbol{n} \tag{2.3}$$

in which $\boldsymbol{f}$ is the curvature vector of the line of force, $H = k_1 + k_2$ the mean curvature of the equipotential surface σ, k_1 and k_2 the principal curvatures of $\sigma, H^* = H + (2\omega^2 - 4\pi G\delta)/g$; we therefore have:

$$\frac{d\boldsymbol{g}}{ds} = -g\left\{\frac{d\boldsymbol{n}}{ds} + \boldsymbol{n}\left[(\boldsymbol{f} - H^*\boldsymbol{n}) \cdot \frac{dP}{ds}\right]\right\} . \tag{2.4'}$$

In vacuo, and in the absence of rotation, (2.4′) reduces to

$$\frac{d\boldsymbol{g}}{ds} = -g\left\{\frac{d\boldsymbol{n}}{ds} + \boldsymbol{n}\left[(\boldsymbol{f} - H\boldsymbol{n}) \cdot \frac{dP}{ds}\right]\right\} \tag{2.4''}$$

which applies, e.g. to the case of inertial space navigation.

The vector field $d\boldsymbol{g}/ds$ as expressed by (2.4″) is called the tidal field acting on the surrounding of P_0.

It should be noted that formulae (2.1), (2.4′) and (2.4″) express $\boldsymbol{g}$ and $d\boldsymbol{g}/ds$ respectively, as the product of the intensity of the acceleration times purely geometrical quantities.

3. The Cardinal and the Canonical Triads

We shall now define at each point P_0 a local orthonormal triad $\mathscr{E}(P_0, \boldsymbol{e}_1, \boldsymbol{e}_2, \boldsymbol{e}_3)$ of mutually orthogonal unit vectors $\boldsymbol{e}_i$ $(i = 1, 2, 3)$ and introduce local Cartesian coordinates x_i based on it. We then have (summation over duplicated indices!):

$$\boldsymbol{e}_i = \frac{\partial P}{\partial x_i} = \operatorname{grad} x_i ; \quad dP = \boldsymbol{e}_i dx_i ; \quad \frac{dP}{ds} = \boldsymbol{e}_i \lambda_i , \quad \lambda_i = \frac{dx_i}{ds} , \tag{3.1'}$$

$$\boldsymbol{g} = W_i \boldsymbol{e}_i , \quad W_i = \frac{\partial W}{\partial x_i} , \tag{3.1''}$$

$$n = \nu_i e_i, \quad f = f_i e_i, \quad \frac{dn}{ds} = \nu_{ij}\lambda_i e_j \tag{3.1'''}$$

$$\frac{dg}{ds} = W_{ij}\lambda_i e_j, \quad W_{ij} = \frac{\partial^2 W}{\partial x_i \partial x_j} = W_{ji} \tag{3.1^{iv}}$$

and equation (2.4′) is now written as

$$W_{ij} = -g[\nu_{ij} + \nu_i(f_j - H^* \nu_j)] \,. \tag{3.2}$$

For an observer connected with the Earth, the natural local triad, intrinsically defined within the field itself, is the canonical triad $\mathscr{C}(P_0, c_1, c_2, c_3)$ defined by the unit vectors c_1 and c_2 of the principal directions of the equipotential surface, and $c_3 = n$. Such a triad is related to the cardinal triad $\mathscr{A}(P_0, a_1, a_2, a_3)$ defined by the horizontal unit vectors a_1 and a_2 in the north and east (cardinal) directions respectively, $a_3 = n$, by simple formulae (Marussi 1951, p. 14).

With respect to the canonical triad $\mathscr{C}$ the matrix W_{ij} takes the simple form

$$W_{ij} \equiv \begin{Vmatrix} -g & 0 & 0 \\ 0 & -g & 0 \\ 0 & 0 & -g \end{Vmatrix} \begin{Vmatrix} k_1 & 0 & f_1 \\ 0 & k_2 & f_2 \\ f_1 & f_2 & -H^* \end{Vmatrix} \tag{3.3}$$

in which k_1 and k_2 are the principal curvatures of the equipotential surface in the directions c_1 and c_2 respectively, and $f_1 = f \cdot c_1 = f \cos\phi$, $f_2 = f \cdot c_2 = f \sin\phi$, ϕ being the angle between c_1 and the direction of the principal normal of the line of force.

4. The Tidal Triad

It is immediately seen from Eq. (3.1^{iv}) that the tidal field at a point $P(x_i)$ in the neighbourhood of P_0 is given by

$$t_P = W_{ij} x_i e_j = \mathrm{grad}\, \frac{1}{2} W_{ij} x_i x_j \,; \tag{4.1}$$

therefore

$$T = \frac{1}{2} W_{ij} x_i x_j \tag{4.2}$$

is the tidal potential acting in the surroundings of P_0.

The surfaces $T =$ constant form a family of concentric homothetic quadrics, the axes of which are the eigenvectors τ_i of the matrix W_{ij}. The orthonormal triad $\mathscr{J}(P_0, \tau_1, \tau_2, \tau_3)$ is the only intrinsic triad recognizable locally with conceivable experiments by an observer in inertial motion around the planet. We shall call it the tidal triad.

In the canonical reference triad $\mathscr{C}$ the tidal potential for an orbiting satellite takes the simple form

$$2T = -g(k_1 x_1^2 + k_2 x_2^2 - H x_3^2 + 2 f_1 x_1 x_3 + 2 f_2 x_2 x_3) \tag{4.3}$$

and since in the gravitational field k_1 and k_2 are always positive, the equipotential surfaces form a family of homothetical hyperboloids of one sheet for $T < 0$ and a family of hyperboloids of two sheets for $T > 0$. The two families are separated by the asymptotic cone $T = 0$. The radial component of the force at a point P near to P_0 is given by $2T/d$, in which d is the distance PP_0. The analogy with the stress field in the classical theory of linear elasticity is evident (Fig. 1).

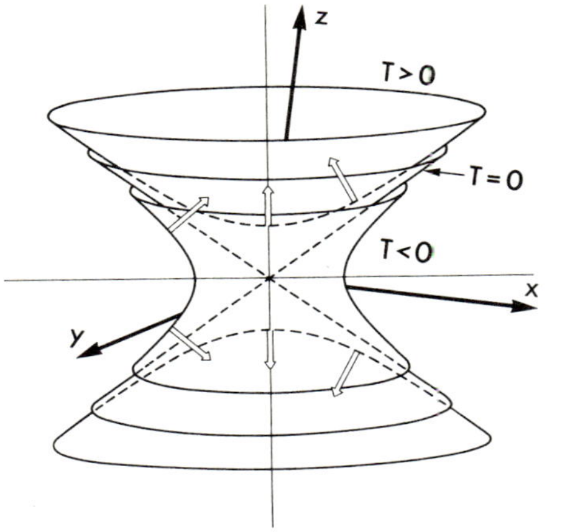

Fig. 1. The equipotential surfaces of the tidal potential

It should be noted that the quadrics (4.3) are equilateral, as a consequence of Laplace's equation being satisfied.

The resultant acceleration at P is given by the sum of the following components:

$$
\begin{aligned}
\boldsymbol{t}(x_1,0,0) &= -g x_1(k_1 \boldsymbol{c}_1 + f_1 \boldsymbol{c}_3)\\
\boldsymbol{t}(0,x_2,0) &= -g x_2(k_2 \boldsymbol{c}_2 + f_2 \boldsymbol{c}_3)\\
\boldsymbol{t}(0,0,x_3) &= -g x_3(\boldsymbol{f} - H\boldsymbol{c}_3) = -g x_3 \operatorname{grad} \ln g\,. \qquad (4.4)
\end{aligned}
$$

The secular equation that furnishes the eigenvalues m of the matrix W_{ij} is written

$$m^3 - (K + C + f^2)m + KH + k_b f^2 = 0 \qquad (4.5)$$

where $K = k_1 k_2$ is the Gaussian, $C = k_1^2 + k_2^2$ Casorati's curvature of the equipotential surface σ, and k_b the normal curvature of the same surface in the direction $\phi + 90°$ of the binormal of the line of force.

Putting

$$\boldsymbol{\tau}_i = \sin\zeta_i \cos\alpha_i \boldsymbol{c}_1 + \sin\zeta_i \sin\alpha_i \boldsymbol{c}_2 + \cos\zeta_i \boldsymbol{c}_3 \qquad (4.6)$$

the directions of the eigenvectors (the tidal axes) are given by

$$
\begin{aligned}
\tan^2\zeta &= \frac{f_1^2}{(m-k_1)^2} + \frac{f_2^2}{(m-k_2)^2}\\
\tan\alpha &= \frac{f_2(m-k_1)}{f_1(m-k_2)}\,. \qquad (4.7)
\end{aligned}
$$

All three roots of the secular equation are real, but they cannot be expressed in closed form by real numbers. Nevertheless, the trigonometric solution of the equation can be easily found.

In the case of rotational symmetry (normal field), for which $k_b = k_2$, the solution can be given in closed form. In this case

$$\left.\begin{aligned} m_1 &= k_2 \\ m_2 &= \frac{-k_2+(H+k_1)\sqrt{1+\eta^2}}{2} \simeq k_1 + \frac{f^2}{H+k_1} \\ m_3 &= \frac{-k_2-(H+k_1)\sqrt{1+\eta^2}}{2} \simeq -H - \frac{f^2}{H+k_1} \end{aligned}\right\} \tag{4.8}$$

$$\eta = \frac{2f}{H+k_1}$$

and the eigenvectors are defined by Eq. (4.7); approximately:

$$\zeta_1 = 90^\circ\,, \qquad \alpha_1 = 90^\circ$$

$$\tan\zeta_2 \simeq \frac{H+k_1}{f}\,, \qquad \alpha_2 = 0$$

$$\tan\zeta_3 \simeq -\frac{f}{H+k_1}\,, \qquad \alpha_3 = 0\,. \tag{4.9}$$

5. The Centrifugal Field

If a centrifugal acceleration characterized by the vector

$$\bar{\omega} = \bar{\omega}_1 c_1 + \bar{\omega}_2 c_2 + \bar{\omega}_3 c_3 \tag{5.1}$$

is superposed in the tidal acceleration, the potential

$$\frac{1}{2}[\bar{\omega} \wedge (P-P_0)]^2 = \frac{\bar{\omega}^2 R^2}{2} \tag{5.2}$$

must be added to T. We have

$$\bar{\omega}^2 R^2 = (\bar{\omega}_2^2+\bar{\omega}_3^2)x_1^2+(\bar{\omega}_1^2+\bar{\omega}_3^2)x_2^2-(\bar{\omega}_1^2+\bar{\omega}_2^2)x_3^2-2(\bar{\omega}_1\bar{\omega}_2x_1x_2+\bar{\omega}_1\bar{\omega}_3x_1x_3 + \bar{\omega}_2\bar{\omega}_3x_2x_3) \tag{5.3}$$

and therefore the total potential is given by

$$2T+\bar{\omega}^2R^2 = (\bar{\omega}_2^2+\bar{\omega}_3^2-gk_1)x_1^2+(\bar{\omega}_1^2+\bar{\omega}_3^2-gk_2)x_2^2+(\bar{\omega}_1^2+\bar{\omega}_2^2+gH)x_3^2 - 2[\bar{\omega}_1\bar{\omega}_2x_1x_2+(\bar{\omega}_1\bar{\omega}_3+gf_1)x_1x_3+(\bar{\omega}_2\bar{\omega}_3+gf_2)x_2x_3]\,. \tag{5.4}$$

For small values of $\bar{\omega}$, the equipotential quadrics are still hyperboloids as before, although no longer equilateral; for larger values of $\bar{\omega}$ they might become cylinders or ellipsoids, depending also upon the direction of $\bar{\omega}$.

6. Reference Frames; Ricci's Coefficients

In order to connect the triads $\mathscr{E}$ referring to different points in space, we note that the integral lines of the differential equation

$$dQ = \boldsymbol{e}_i ds_i \quad (i = 1, 2, 3) \tag{6.1}$$

form a triple system $E_{(k)}$ of mutually orthogonal congruences. We may assume the lines of the system of congruences $E_{(k)}$ as coordinate lines and use the length s_i of the corresponding arcs as coordinates, thus defining a frame of reference.

In this frame, two neighbouring triads are connected by Ricci's (1895) coefficients of rotation (summation over duplicated indices implied) defined by the following equations:

$$\gamma_{ijk} = \frac{\partial \boldsymbol{e}_i}{\partial s_k} \cdot \boldsymbol{e}_j = \frac{\partial \boldsymbol{e}_j}{\partial s_k} \cdot \boldsymbol{e}_i ; \qquad \frac{\partial \boldsymbol{e}_i}{\partial s_k} = \gamma_{ijk} \boldsymbol{e}_j ; \qquad \gamma_{ijk} = -\gamma_{jik} . \tag{6.2}$$

The connection between a given triad and the neighbouring one can be also described as a rigid body motion; we shall therefore put

$$\frac{\partial \boldsymbol{e}_i}{\partial s_k} = \gamma_{ijk} \boldsymbol{e}_j = \boldsymbol{\pi}_k \wedge \boldsymbol{e}_i \tag{6.3}$$

in which $\boldsymbol{\pi}_k$ is Pensa's (1913 – 14) rotation vector which depends upon the direction connecting the two triads. We shall also put

$$\boldsymbol{\pi}_k = \pi_{sk} \boldsymbol{e}_s ; \tag{6.4}$$

π_{sk} are the coefficients of Pensa by means of which Ricci's coefficients of rotation can be expressed as follows:

$$\gamma_{ijk} = e_{ijs} \pi_{ks} , \qquad \pi_{ks} = \frac{1}{2} e_{ijs} \gamma_{ijk} . \tag{6.5}$$

Here e_{ijs} is equal to $+1$ if the permutation of the indices is even, and to -1 if the permutation is odd: it is zero if two or more indices are equal.

By the definition of the rotation operator we may write

$$\operatorname{curl} \boldsymbol{e}_k = \frac{1}{2} e_{ijs} \Omega_{kji} \boldsymbol{e}_s \tag{6.6}$$

in which

$$\Omega_{kji} = \gamma_{kji} - \gamma_{kij} , \qquad \Omega_{kji} = -\Omega_{kij} \tag{6.7}$$

are called the objects of anholonomy associated with the congruence $E_{(k)}$ of the reference frame. We have explicitly

$$\left.\begin{aligned} \operatorname{curl} \boldsymbol{e}_1 &= \Omega_{132} \boldsymbol{e}_1 + \Omega_{113} \boldsymbol{e}_2 + \Omega_{121} \boldsymbol{e}_3 \\ \operatorname{curl} \boldsymbol{e}_2 &= \Omega_{232} \boldsymbol{e}_1 + \Omega_{213} \boldsymbol{e}_2 + \Omega_{221} \boldsymbol{e}_3 \\ \operatorname{curl} \boldsymbol{e}_3 &= \Omega_{332} \boldsymbol{e}_1 + \Omega_{313} \boldsymbol{e}_2 + \Omega_{321} \boldsymbol{e}_3 \end{aligned}\right\} \tag{6.8}$$

or, in terms of Pensa's coefficients

$$\operatorname{curl} \boldsymbol{e}_1 = -(\pi_{22}+\pi_{33})\boldsymbol{e}_1+\pi_{12}\boldsymbol{e}_2+\pi_{13}\boldsymbol{e}_3$$
$$\operatorname{curl} \boldsymbol{e}_2 = \pi_{21}\boldsymbol{e}_1-(\pi_{11}+\pi_{33})\boldsymbol{e}_2+\pi_{23}\boldsymbol{e}_3$$
$$\operatorname{curl} \boldsymbol{e}_3 = \pi_{31}\boldsymbol{e}_1+\pi_{32}\boldsymbol{e}_2-(\pi_{11}+\pi_{22})\boldsymbol{e}_3 . \tag{6.9}$$

We also have by the definition of the divergence operator

$$\operatorname{div} \boldsymbol{e}_1 = \pi_{23}-\pi_{32}$$
$$\operatorname{div} \boldsymbol{e}_2 = \pi_{31}-\pi_{13}$$
$$\operatorname{div} \boldsymbol{e}_3 = \pi_{12}-\pi_{21} . \tag{6.10}$$

If for a given k the following equation holds

$$\boldsymbol{e}_k \cdot \operatorname{curl} \boldsymbol{e}_k = \Omega_{k,k+1,k+2} = 0 \tag{6.11}$$

(3 must be subtracted from the indices if they exceed 3), the congruence $E_{(k)}$ of the reference frame is normal, i.e., its lines intersect at right angles a family of surfaces $\Sigma_{(k)}$. In this case an integrating factor $\phi_{(k)}$ exists, such that

$$\operatorname{curl} \phi_{(k)} \cdot \boldsymbol{e}_k = \phi_{(k)} \operatorname{curl} \boldsymbol{e}_k + \operatorname{grad} \phi_{(k)} \wedge \boldsymbol{e}_k = 0 \tag{6.12}$$

and therefore

$$\phi_{(k)}\boldsymbol{e}_k = \operatorname{grad} \psi_{(k)} . \tag{6.13}$$

If Eq. (6.11) holds for any k, all the lines of the reference frame are the intersections of pairs of surfaces belonging to a triple system of families of which they are the lines of curvature. A system of coordinates can be associated in this case and only in this case to the reference frame, that is therefore called an holonomic frame of reference.

In general Eq. (6.11) is not satisfied for all k, and the frame or reference is anholonomic. We have in this case that the second derivatives with respect to the arcs s_i of any scalar function ϕ are not commutable:

$$\left(\frac{\partial^2}{\partial s_j \, \partial s_k}-\frac{\partial^2}{\partial s_k \, \partial s_j}\right)\phi = \Omega_{ikj}\frac{\partial \phi}{\partial s_i} . \tag{6.14}$$

7. The Cardinal Frames

If one and only one of the congruences, say the congruence E_3 is normal, the following relations hold.

$$\Omega_{3jk} = 0 \qquad (j,k = 1,2)$$
$$\Omega_{ijk} \neq 0 \qquad (i \neq 3, i \neq j) . \tag{7.1}$$

Such is the case for the congruence F_3 of the lines of force in the gravity field of a planet, since these lines intersect at right angles the equipotential surfaces $W=$ constant. The cardinal system of congruences formed by the north and the east lines on these surfaces (Marussi 1951, p. 18), can then be adopted as associated congruences in the reference frame. Any other pair of mutually orthogonal congruences belonging to the family of equipotential surfaces can be related to the cardinal system by means of a parameter θ, e.g., the angle (azimuth) formed by the lines of one of two congruences with the north lines.

We may dispose of such parameter to satisfy the following further condition

$$\gamma_{ijk} = -\gamma_{ikj} \quad (i = 3, j \neq k, j, k = 1, 2) . \tag{7.2}$$

The system of congruences satisfying both the conditions (7.1) and (7.2) is called a canonical system.

From Eq. (7.1) we have in this case

$$\gamma_{321} - \gamma_{312} = 0 \tag{7.1'}$$

and from Eqs. (6.2) and (7.2)

$$\gamma_{321} + \gamma_{312} = 0 \tag{7.2'}$$

and therefore

$$\gamma_{321} = \gamma_{312} = 0 . \tag{7.3}$$

It follows that

$$\frac{\partial \boldsymbol{e}_3}{\partial s_1} = \gamma_{311} \boldsymbol{e}_1 , \qquad \frac{\partial \boldsymbol{e}_3}{\partial s_2} = \gamma_{322} \boldsymbol{e}_2 , \tag{7.4}$$

i.e., in the canonical system the lines of the associated congruences are the lines of curvature and γ_{311} and γ_{322} are the respective principal curvatures of the equipotential surfaces. The local reference triad is therefore the canonical triad $\mathscr{C}$ described in Section 3.

Since we shall deal with intrinsic properties of the gravitational field only, we shall refer in the following to such a canonical frame of reference.

Whenever necessary the angle θ can be found by the procedure indicated in (Marussi 1951, p. 17).

8. The Canonical Frame

In the canonical case, the coefficients of Pensa, the coefficients of connection and the objects of anholonomy are given by the following matrices in which the first running index refers to the rows and the second to the columns:

$$\pi_{ij} \equiv \begin{Vmatrix} 0 & k_1 & k_{g1} \\ -k_2 & 0 & k_{g2} \\ -f_2 & f_1 & -\tau_g^f \end{Vmatrix} \tag{8.1}$$

$$\gamma_{1ij} \equiv \begin{Vmatrix} 0 & 0 & 0 \\ k_{g1} & k_{g2} & -\tau_g^f \\ -k_1 & 0 & -f_1 \end{Vmatrix} \qquad \Omega_{1ij} \equiv \begin{Vmatrix} 0 & -k_{g1} & k_1 \\ k_{g1} & 0 & -\tau_g^f \\ -k_1 & \tau_g^f & 0 \end{Vmatrix} \tag{8.2'}$$

$$\gamma_{2ij} \equiv \begin{Vmatrix} -k_{g1} & -k_{g2} & \tau_g^f \\ 0 & 0 & 0 \\ 0 & -k_2 & -f_2 \end{Vmatrix} \qquad \Omega_{2ij} \equiv \begin{Vmatrix} 0 & -k_{g2} & \tau_g^f \\ k_{g2} & 0 & k_2 \\ -\tau_g^f & -k_2 & 0 \end{Vmatrix} \tag{8.2''}$$

$$\gamma_{3ij} \equiv \begin{Vmatrix} k_1 & 0 & f_1 \\ 0 & k_2 & f_2 \\ 0 & 0 & 0 \end{Vmatrix} \qquad \Omega_{3ij} \equiv \begin{Vmatrix} 0 & 0 & f_1 \\ 0 & 0 & f_2 \\ -f_1 & -f_2 & 0 \end{Vmatrix}. \tag{8.2'''}$$

Here k_{g1} and k_{g2} are the geodetic curvatures of the lines of curvature of the equipotential surfaces in the directions c_1 and c_2 respectively, τ^f the torsion of the lines of force, and $\tau_g^f = \tau^f + d\phi/ds_3$ the geodetic torsion of the congruence of the lines of force.

It is also easily seen that the following relations hold:

$$\left.\begin{aligned} \mathrm{curl}\,\boldsymbol{c}_1 &= \tau_g^f \boldsymbol{c}_1 + k_1 \boldsymbol{c}_2 + k_{g1} \boldsymbol{c}_3\,; \quad \boldsymbol{c}_1 \cdot \mathrm{curl}\,\boldsymbol{c}_1 = \tau_g^f \\ \mathrm{curl}\,\boldsymbol{c}_2 &= -k_2 \boldsymbol{c}_1 + \tau_g^f \boldsymbol{c}_2 + k_{g2} \boldsymbol{c}_3\,; \quad \boldsymbol{c}_2 \cdot \mathrm{curl}\,\boldsymbol{c}_2 = \tau_g^f \\ \mathrm{curl}\,\boldsymbol{c}_3 &= -f_2 \boldsymbol{c}_1 + f_1 \boldsymbol{c}_2\,; \quad \boldsymbol{c}_3 \cdot \mathrm{curl}\,\boldsymbol{c}_3 = 0 \end{aligned}\right\} \tag{8.3}$$

and

$$\left.\begin{aligned} \mathrm{div}\,\boldsymbol{c}_1 &= k_{g2} - f_1 \\ \mathrm{div}\,\boldsymbol{c}_2 &= k_{g1} + f_2 \\ \mathrm{div}\,\boldsymbol{c}_3 &= H \end{aligned}\right\}. \tag{8.4}$$

9. The Tidal Frame

As already pointed out in Section 4, in the case of inertial navigation only the tidal axes τ_i can be recognized with conceivable experiments; the triple orthogonal system of congruences defined by the equation

$$d\phi = \boldsymbol{\tau}_i \, d\sigma_i \tag{9.1}$$

represents therefore the natural reference frame of coordinate lines that should be considered in this case.

Since the tidal axes $\boldsymbol{\tau}_i$ are related to the canonical axes $\boldsymbol{c}_i$ by Eq. (4.6), the computation of Ricci's coefficients of rotation and associated quantities can be obtained in principle, although not explicitly, by differentiating this equation.

Bibliography

Marussi A (1951) Fondamenti di geodesia intrinseca. Comm Geodet Ital Mem no 7 (Chap. I.2, this Vol.)

Pensa A (1913 – 14) Sulla risoluzione di equazioni vettoriali ed omografiche. Atti R Accad Sci Torino 49:987 – 1012

Ricci G (1895) Dei sistemi di congruenze ortogonali in una varietà qualunque. Atti R Accad Lincei, anno CCXCII: Ser 5, Mem II, 276 – 322

II Structure of the Gravity Field and Laplace's Equation

1 On the Curvature and Torsion of the Gravity Field[1]

1. Considering two points P_1 and P_2 of the Earth's external gravity field, sufficiently close to each other, and the two (vertical) versors $\boldsymbol{n}_1$ and $\boldsymbol{n}_2$, directed upwards which are tangent to the lines of force at P_1 and P_2, then the axes of two such versors will not, in general, be coplanar. When the two points lie on the same equipotential surface Σ, such a defect of coplanarity appears, as is well known, as the presence of the geodesic torsion of Σ in the direction P_1P_2. If we put $P_2 - P_1 = s\boldsymbol{t}$, where $\boldsymbol{t}$ is a unit vector, and s the length of the segment P_1P_2 which we suppose to be small to a first order, we then have from the generalized formulae of Frenet

$$\frac{\boldsymbol{n}_2 - \boldsymbol{n}_1}{s} = \varkappa_\alpha \boldsymbol{t} + \tau_\alpha \boldsymbol{b}\,, \tag{1}$$

where $\boldsymbol{b}$ is a versor such that $(\boldsymbol{t}, \boldsymbol{b}, \boldsymbol{n})$ form a positive right-handed triad, $\varkappa_\alpha$ is the first curvature or flexion of the surface in the direction $\boldsymbol{t}$ considered, and τ_α the geodesic torsion of Σ in the same direction. The index α characterizes the direction $\boldsymbol{t}$, for the reason which will later be apparent.

But, in the same way, considering the two points P_1 and P_2 no longer on the same equipotential surface, but still in space, it will be possible to define what we shall call by analogy the curvature and torsion of the gravity field, in the direction characterized by the versor $\boldsymbol{a}$ such that $P_2 - P_1 = s\boldsymbol{a}$.

We propose to examine these characteristics, showing in particular how the second (the torsion) can be completely determined by means of the Eötvös balance of the second kind, along with the measurement of gravity, and then investigating the directions of zero curvature and torsion of the field.

2. For this, let us put

$$\boldsymbol{a} = \sin z \cdot \boldsymbol{t} + \cos z \cdot \boldsymbol{n}\,, \tag{2}$$

where z then represents the zenith distance of $\boldsymbol{a}$, and $\boldsymbol{t}$ the direction of its projection on Σ, the equipotential surface passing through P_1.

For the surface displacement from P_1 to P_2', the foot-point of the line of force through P_2, a displacement whose magnitude is $s \sin z$, we have from the preceding formula

$$\frac{\boldsymbol{n}_2' - \boldsymbol{n}_1}{s \sin z} = \varkappa_\alpha \boldsymbol{t} + \tau_\alpha \boldsymbol{b}\,; \tag{3}$$

[1] Originally published as: Marussi A (1952) Sulla curvatura e torsione del campo di gravità. Ann Geofis 5: no 2

while for the displacement along the line of force, whose magnitude is $s \cos z$, we have the well known result

$$\frac{n_2 - n_2'}{s \cos z} = \mathrm{grad}_\Sigma \log g = f\boldsymbol{B} \ , \tag{4}$$

where f is the curvature (essentially positive) of the line of force, and $\boldsymbol{B}$ its principal normal; f also represents obviously the modulus of the horizontal component of the logarithmic gradient of gravity.

Decomposing $\boldsymbol{B}$ along $\boldsymbol{t}$ and $\boldsymbol{b}$ we have

$$\boldsymbol{B} = \cos\psi \cdot \boldsymbol{t} + \sin\psi \cdot \boldsymbol{b} \tag{5}$$

and finally, adding Eqs. (3) and (4)

$$\frac{n_2 - n_1}{s} = (\varkappa_\alpha \sin z + f \cos z \cos\psi)\boldsymbol{t} + (\tau_\alpha \sin z + f \cos z \sin\psi)\boldsymbol{b} \ . \tag{6}$$

We thus define by analogy as the *curvature* C_a and the *torsion* T_a of the field in the direction $\boldsymbol{a}$

$$C_a = \varkappa_\alpha \sin z + f \cos z \cos\psi$$

$$T_a = \tau_\alpha \sin z + f \cos z \sin\psi \ . \tag{7}$$

Moreover, these formulae can be put into a more expressive form, with the involvement of only absolute elements. Denoting precisely by α the bearing of $\boldsymbol{t}$ with respect to $\boldsymbol{c}_1$, where $\boldsymbol{c}_1$ is the principal direction of curvature $\varkappa_1$, we have from Euler's formula

$$\varkappa_\alpha = \varkappa_1 \cos^2\alpha + \varkappa_2 \sin^2\alpha \ ; \tag{8}$$

and further

$$\tau_\alpha = \frac{\sin 2\alpha}{2}(\varkappa_2 - \varkappa_1)$$

$$\psi = \vartheta - \alpha \tag{9}$$

where ϑ is the horizontal angle between $\boldsymbol{B}$ and $\boldsymbol{c}_1$. Hence

$$C_a = (\varkappa_1 \cos^2\alpha + \varkappa_2 \sin^2\alpha) \sin z + f \cos z \cos(\vartheta - \alpha)$$

$$T_a = \frac{\sin 2\alpha \sin z}{2}(\varkappa_2 - \varkappa_1) + f \cos z \sin(\vartheta - \alpha) \ . \tag{10}$$

These are the desired expressions.

3. We can now ask the questions: are there directions in space along which the field curvatures are zero? or along which the torsion is zero?, or finally, are there directions along which both the curvature and the torsion are zero?

Replying to the first question, and putting $C_a = 0$, we have the equation

$$\mathrm{tg}\, z = \frac{-f \cos(\vartheta - \alpha)}{\varkappa_1 \cos^2\alpha + \varkappa_2 \sin^2\alpha} \ , \tag{11}$$

which represents a system ∞^1 of directions. In spherical representation these trace out a curve whose form we can readily imagine by noting that the minimum value of the denominator is equal to the minimum $\varkappa$ of $\varkappa_1$ and $\varkappa_2$; and the maximum value of the numerator is equal to f. Hence the tangent of z is always kept less than the ratio $f/\varkappa$ and thus, if we wish, lower than the ratio of the maximum radius of the equipotential surface to the radius of first curvature of the line of force. Such a ratio is usually quite small; for the normal field at 45° it is about 0.0053; and hence, in this case, the zenith distance of the line of zero curvature is kept at less than 18′.

Setting z equal to zero, α tends to $\vartheta \pm 90°$, and thus for $z = 0$ the curve allows a tangent of bearing $\vartheta \pm 90°$, hence perpendicular to the horizontal component of the gravity gradient. The curve also lies entirely on one side of this tangent, and in particular on the side opposite to that toward which the above component is directed. In the particular case in which $\varkappa_1 = \varkappa_2$, the equation represents, on the plane tangent to the sphere of directions at the zenith point, a small circle (see Figs. 1, 2 and 3, which illustrate three distinct cases, where the dimensions of the curve have been greatly exaggerated).

4. The directions of zero torsion are obtained from the equation

$$T_a = -(\varkappa_1 - \varkappa_2)\sin\alpha\cos\alpha\sin z + f\cos z\sin(\vartheta - \alpha) = 0\,; \tag{12}$$

leaving aside the cases $\vartheta = 0$, $\vartheta = 90°$, as well as $\varkappa_1 = \varkappa_2$, which will be examined later, this equation gives

$$\operatorname{tg} z = \frac{f}{\varkappa_1 - \varkappa_2}\left(\frac{\sin\vartheta}{\sin\alpha} - \frac{\cos\vartheta}{\cos\alpha}\right). \tag{13}$$

For $\vartheta = 0$, $\vartheta = 90°$, $\varkappa_1 = \varkappa_2$ we have instead the solutions

$$\text{for}\quad \vartheta = 0: \quad \alpha = 0\,, \quad \operatorname{tg} z = \frac{f}{(\varkappa_1 - \varkappa_2)\cos\alpha}$$

$$\text{for}\quad \vartheta = 90°: \quad \alpha = 90°\,, \quad \operatorname{tg} z = \frac{f}{(\varkappa_1 - \varkappa_2)\sin\alpha}$$

$$\text{for}\quad \varkappa_1 = \varkappa_2 \quad \alpha = \vartheta\,, \quad z = 90°$$

which represent: the first, the vertical plane containing $\boldsymbol{c}_1$ and the plane normal to it which forms with the horizontal an angle whose tangent is equal to $(\varkappa_1 - \varkappa_2)/f$ (see Fig. 1 which shows the normal case at latitude 45°; here $f = 0.830$, $\varkappa_1 - \varkappa_2 = 0.530$, $(\varkappa_1 - \varkappa_2)/f = 0.639$, the curvatures being measured in 10^{-6} km^{-1}); the second, the vertical plane containing $\boldsymbol{c}_2$ an the plane perpendicular to it inclined as was seen above; finally the third, the horizon and the vertical plane which contains the horizontal component of the gravimetric gradient.

In the general case synthesized from Eq. (13), the curve in spherical representation has the form shown in stereographic projection (restricted to the upper half of the sphere) in Figs. 2 and 3, where are also given the parameters adopted. The branch of the curve which passes through the zenith has here as tangent the vertical of bearing ϑ, as can be readily seen.

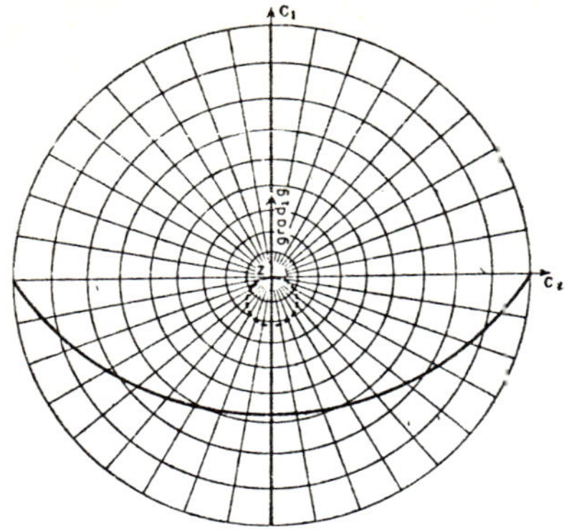

Fig. 1. Latitude 45°; normal case

$\vartheta = 0\,, \quad \varkappa_1 = 157.1\,; \quad f = 0.83$

$\varkappa_1 - \varkappa_2 = 0.52\,; \quad \dfrac{f}{\varkappa_1 - \varkappa_2} = 1.566$

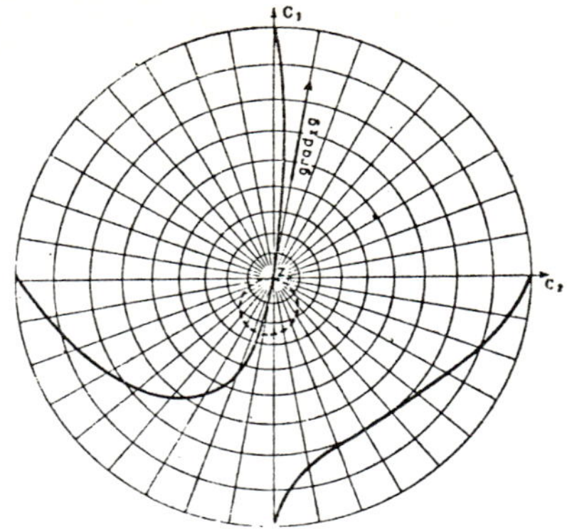

Fig. 2. Latitude 45°; $\vartheta = 10°$

$\varkappa_1 = 157.1\,; \quad f = 1.06\,; \quad \varkappa_1 - \varkappa_2 = 0.53$

$\dfrac{f}{\varkappa_1 - \varkappa_2} = 2.00$

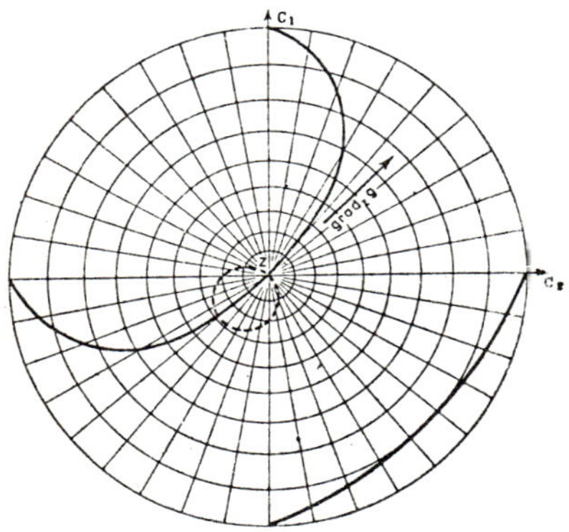

Fig. 3. Latitude 45°; $\vartheta = 45°$

$\varkappa_1 = 157.1\,; \quad f = 1.06\,; \quad \varkappa_1 - \varkappa_2 = 0.53$

$\dfrac{f}{\varkappa_1 - \varkappa_2} = 2.00$

The three figures show, in a zenithal stereographic projection, the loci of zero curvature (*broken line*), and those of zero torsion (*full line*) of the gravity field, traced out on a unit sphere of directions. *Outer circle* horizon; *Z* zenith; c_1 and c_2 principal directions of curvature of the equipotential surface. The dimensions of the loci of zero curvature are greatly exaggerated

5. Replying to the third question, and setting aside of the singular case of the zenith, where curvature and torsion lose their meaning, it appears that only one direction has both torsion and curvature simultaneously zero; along this direction the verticals keep parallel to each other, and this therefore represents the *isozenithal* line.

This can also be obtained by equating Eqs. (11) and (13); which gives

$$\frac{\operatorname{tg}\alpha - \operatorname{tg}\vartheta}{\operatorname{tg}\alpha} = (\varkappa_1 - \varkappa_2)\,\frac{1 + \operatorname{tg}\alpha\operatorname{tg}\vartheta}{\varkappa_1 + \varkappa_2\operatorname{tg}^2\alpha} \tag{14}$$

an equation which is satisfied by putting

$$\operatorname{tg} \alpha = \frac{\varkappa_1}{\varkappa_2} \operatorname{tg} \vartheta . \tag{15}$$

Hence this provides the bearing of the isozenithal line.

6. For the practical determination of the curvature and torsion of the gravity field, which can be viewed either from a regional or from a local point of view, the methods of geodetic astronomy should provide, at least in theory, for the first, and methods based on the use of the torsion balance for the second. It should, however, be noted straight away that while the Eötvös balance of the second type, together with the determination of gravity, completely defines the torsion of the field, it does not define the curvature, at least not without making use of well-known integration procedures.

In the astronomic-geodetic method, we determine n_1 and n_2 at P_1 and P_2 by astronomical measurements of latitude and longitude, the astronomical azimuth A and the zenith distance z from P_1 to P_2, and the actual distance s between the two points (the latter can be had to a good approximation by triangulation, or more rigorously by a geodimeter), and thus arrive at the "regional" values C_a and T_a of curvature and torsion in the direction A considered. Putting then

$$A = \alpha + v ; \qquad \chi = \vartheta + v \tag{16}$$

where v and χ are respectively the astronomical azimuths of the direction of curvature $\varkappa_1$ and of the surface gradient of gravity, we have the two equations

$$C_a = [\varkappa_1 \cos^2 (A - v) + \varkappa_2 \sin^2 (A - v)] \sin z + f \cos z \cos (\chi - A)$$

$$T_a = \frac{\sin^2 (A - v) \sin z}{2} (\varkappa_2 - \varkappa_1) + f \cos z \sin (\chi - A) \tag{17}$$

in the five unknows $\varkappa_1, \varkappa_2, f, v, \chi$; three observations of this type appropriately chosen and made around the point P_1 will allow us in theory to determine all the unknowns sought.

For the local determination by the Eötvös balance, it is sufficient to remember that the output of this instrument is a reflection of the non-coplanarity of the lines of action of gravity on the extremities of the balance-arm; and more precisely, the twisting moment with respect to the vertical is given, in an ideal balance formed by two equal point masses m arranged along the versor $\boldsymbol{a}$ previously considered, and separately by a distance a, by

$$M_a = m a^2 g T_a \sin z . \tag{18}$$

If then g is known, and the moment is measured with the proper instrumental procedures, then T_a is completely determined; four observations (to which must be added a fifth for well-known instrumental reasons) will determine the quantities $\varkappa_1 - \varkappa_2, f, v$ and χ, and from these the torsion in any given direction; and thus we can also find the loci of directions of zero torsion considered above.

2 Generalization of Dalby's Theorem for Any Surface Whatever[1]

1. In my previous communication presented to the IX Assembly of the International Association of Geodesy gathered at Brussels in 1951 (Chap. IV.1, this Vol.), I recalled the absolute character inherent in Laplace's equation, which connects the deviations of the vertical in longitude and in azimuth. This arises in particular from the fact that the equation permits us to obtain, according to the case, the astronomical longitudes from the astronomical azimuths, or vice versa, without the ellipsoid used in the definition leaving the slightest trace in the result.

From this point of view Laplace's equation is equivalent to Dalby's theorem, famous at one time for the determination of the differences of astronomical longitudes from the differences of astronomical azimuths, and which by comparison with the preceding equation of Laplace has the advantage of being able to be stated without recourse to an intermediate surface of representation such as the ellipsoid. The very great value of Dalby's theorem in this latter respect has already been fully realized by Ivory (1828), as I have had occasion to recall (Marussi 1951).

Ivory has quite correctly suggested that in no case should one deal with the geodesic joining two station points, whether on the surface of calculation, or worse, on that of representation, since this geodesic is a purely conventional entity, and useful for calculation, but inaccessible to direct measurement. Instead, one should deal with the normal sections, the only ones concerned in the operations of measurement of azimuth. From this viewpoint Dalby's theorem represents the absolute enunciation of Laplace's formula, avoiding any conventional surface or line.

In spite of its exceptional heuristic importance, Dalby's theorem has been considered, as far as I know, only in the case of an ellipsoidal surface. The interest which it sustained in the past, when there was no telegraph for the determination of longitudes, has been lost, and its present modest role is reduced to that of providing one of many methods for computing the reciprocal azimuths on the ellipsoid, for the solution of the second fundamental problem of Geodesy.

In the present chapter I propose, however, to illustrate this fundamental theorem, and to generalize it for any given surface – which in particular could be the surface of the geoid – with the intention that it might be substituted for Laplace's Theorem, when we do not wish to have recourse to the intermediary of a conventional and arbitrary reference surface.

---

[1] Originally published as: Marussi A (1952) Generalizzazione del teorema di Dalby per una superficie qualunque. Festschr Eduard Dolezal

2. The classical Dalby's theorem can be stated in the following manner: "given two points P_1 and P_2 on the ellipsoid of rotation, and the reciprocal normal planes at these points, each containing the other point; considering the spherical images n_1 and n_2 of these same two points, and the spherical triangle formed by these and by the pole; the convergence of the above-mentioned normal planes and the convergence of the corresponding tangents to the great circle joining n_1 and n_2 on the sphere, differ from each other by small quantities of the fifth order in the square of the eccentricity e, and in the difference between the latitudes ϕ and the longitudes λ of the two points".

Specifically denoting by α_1 and α_2 the (reciprocal) azimuths of the normal planes Π_1 and Π_2, by α_1' and α_2' the (reciprocal) azimuths of the great circle on the sphere, we have, to a first approximation (in the case of the ellipsoid),

$$\alpha_1 - \alpha_2 - (\alpha_1' - \alpha_2') = \eta_1 - \eta_2 = \frac{e^4 (\Delta\varphi)^2 \Delta\lambda \sin 2\varphi \cos^3\varphi}{8}. \tag{1}$$

In the general case of any surface Σ, by representing also on the spherical image the direction C of the line joining P_1P_2 which is common to the normal planes Π_1 and Π_2, we see that this same difference is due to the presence of the angles η_1 and η_2, and to the fact that these in general differ between themselves. Now it is easy to see also that the angles η_1 and η_2 are generated by the non-coplanarity of Π_1 and Π_2, which form between them the dihedral angle denoted by ε; and this angle is due to the presence of torsion of the surface in the direction P_1P_2. They then differ from each other in general, since the zenith distances ζ_1 and ζ_2, are not equal to each other; and this is due to the fact that the curvature of Σ along P_1P_2 is not, in general, constant.

It should thus be expected, from what has been discussed, that the principal part of Dalby's term should turn out to be proportional to the product of the torsion τ by the derivative of the first curvature $\varkappa'$ in the direction considered. This we propose to demonstrate, finding also the value of such a term with all the approximation needed for practical applications.

Once Dalby's general term is known, the relationship between the astronomical azimuths of the normal sections, and astronomical longitudes, will be completely determined from the spherical triangle in the image of directions, which is immediately soluble with the usual analogies of Napier.

3. To do this, we shall avail ourselves of the absolute methods of the vector calculus, which allow us to use only intrinsic elements of the surface, without introducing arbitrary reference systems. On our surface the points P_1 and P_2 will be connected by means of geodetic polar coordinates, having the pole at the mean point P_0 of the geodesic which joins the said two points, so as to confer the maximum of symmetry to the formulae.

It remains to be understood that all the curvatures of flexion and of torsion which need to be considered, and their directions, are referred to the point P_0 and to the direction of the geodesic which joins the two given points, in the direction from P_1 to P_2. We shall denote by s the length of the arc of geodesic between P_1 and P_2, and by σ the length of the straight-line chord which joins them in space.

Supposing that Σ has everywhere a positive total curvature, we shall denote by $\boldsymbol{t}$, $\boldsymbol{n}$, $\boldsymbol{b}$ the versors of the tangent, the external normal to Σ, and the binormal to our geodesic at P_0, so that the trihedron $(\boldsymbol{t}, \boldsymbol{n}, \boldsymbol{b})$ will be positive. We shall then have, for example,

$$P_2 = P_0 + \frac{s}{2}\left(\frac{dP}{ds}\right)_0 + \frac{s^2}{8}\left(\frac{d^2P}{ds^2}\right)_0 + \ldots \tag{2}$$

whence noting that $\dfrac{dP}{ds} = \boldsymbol{t}$, and from Frenet's formulae

$$\begin{aligned} \frac{d^2P}{ds^2} &= \frac{d\boldsymbol{t}}{ds} = \varkappa\boldsymbol{n} \\ \frac{d^3P}{ds^3} &= \frac{d^2\boldsymbol{t}}{ds^2} = -\tau\boldsymbol{t} + \varkappa'\boldsymbol{n} - \varkappa\boldsymbol{b} \qquad \text{etc.} \end{aligned} \tag{3}$$

$\varkappa$ and τ being respectively the normal curvature and the torsion of Σ at P_0 along the direction $\boldsymbol{t}$. We thus readily obtain, by taking differences between the analogous expressions for P_2 and P_1,

$$P_2 - P_1 = s\left(1 - \frac{s^2}{24}\varkappa^2\right)\boldsymbol{t} + \frac{s^3}{24}\varkappa'\boldsymbol{n} + \frac{s^3}{24}\varkappa\tau\boldsymbol{b} + \ldots \tag{4}$$

We note that the normal $\boldsymbol{n}$ is directed positively towards the convexity of the geodesic, and $\varkappa$ is thus an essentially negative quantity, while τ instead has a positive sign, since the fundamental trihedron rotates from P_1 towards P_2 in a clockwise sense for an observer looking along $\boldsymbol{t}$.

Similarly, for the normal $\boldsymbol{n}_2$ to Σ at P_2 we have

$$\boldsymbol{n}_2 = \boldsymbol{n}_0 + \frac{s}{2}\left(\frac{d\boldsymbol{n}}{ds}\right)_0 + \frac{s^2}{8}\left(\frac{d^2\boldsymbol{n}}{ds^2}\right)_0 + \ldots \tag{5}$$

and from the same formulae of Frenet we obtain

$$\begin{aligned} \boldsymbol{n}_1 = {} & \frac{s}{2}\left[\varkappa - \frac{s}{4}\varkappa' - \frac{s^2}{24}(\varkappa\omega^2 - \varkappa'')\right]\boldsymbol{t} + \left[1 - \frac{s^2}{8}\omega\left(\omega - \frac{s}{2}\omega'\right)\right]\boldsymbol{n} \\ & - \frac{s}{2}\left[\tau - \frac{s}{4}\tau' + \frac{s^2}{24}(\tau'' - \tau\omega^2)\right]\boldsymbol{b} \\ \boldsymbol{n}_2 = {} & -\frac{s}{2}\left[\varkappa + \frac{s}{4}\varkappa' - \frac{s^2}{24}(\varkappa\omega^2 - \varkappa'')\right]\boldsymbol{t} + \left[1 - \frac{s^2}{8}\omega\left(\omega - \frac{s}{2}\omega'\right)\right]\boldsymbol{n} \\ & + \frac{s}{2}\left[\tau + \frac{s}{4}\tau' + \frac{s^2}{24}(\tau'' - \tau\omega^2)\right]\boldsymbol{b} \end{aligned} \tag{6}$$

in which $\omega^2 = \varkappa^2 + \tau^2$ is the square of the kinematic vector of Darboux.

Again we see that for the reciprocal zenith distances ζ_1 and ζ_2 for P_2P_1 we have respectively

$$\sigma\cos\zeta_1 = (P_2-P_1)\times\boldsymbol{n}_1\,; \qquad \sigma\cos\zeta_2 = -(P_2-P_1)\times\boldsymbol{n}_2 \tag{7}$$

whence

$$\sigma\cos\zeta_1 = \frac{s^2}{2}\left[\varkappa-\frac{s}{6}\varkappa'-\frac{s^2}{12}\left(\varkappa\omega^2-\frac{\varkappa''}{2}\right)\right]$$

$$\sigma\cos\zeta_2 = \frac{s^2}{2}\left[\varkappa+\frac{s}{6}\varkappa'-\frac{s^2}{12}\left(\varkappa\omega^2-\frac{\varkappa''}{2}\right)\right] \tag{8}$$

and again

$$\sigma\sin\zeta_1 = s\left(1-\frac{s^2\varkappa^2}{6}+\frac{s^3\varkappa\varkappa'}{24}\right)$$

$$\sigma\sin\zeta_2 = s\left(1-\frac{s^2\varkappa^2}{6}-\frac{s^3\varkappa\varkappa'}{24}\right). \tag{9}$$

4. Given all this, and denoting respectively by $\boldsymbol{B}_1$ and $\boldsymbol{B}_2$ the versors perpendicular to the normal planes Π_1 and Π_2 considered in Dalby's theorem, which are oriented in the sense that results from the following formulae, we shall have

$$\boldsymbol{B}_1 = \frac{(P_2-P_1)\wedge\boldsymbol{n}_1}{\sigma\sin\zeta_1}\,; \qquad \boldsymbol{B}_2 = \frac{(P_2-P_1)\wedge\boldsymbol{n}_2}{\sigma\sin\zeta_2} \tag{10}$$

and the angle ε between the two forementioned planes (or between the versors perpendicular to them) will be given by

$$\sin\varepsilon = \operatorname{mod}(\boldsymbol{B}_1\wedge\boldsymbol{B}_2) \tag{11}$$

or also

$$\sigma\sin\varepsilon = \boldsymbol{B}_1\wedge\boldsymbol{B}_2\times(P_2-P_1)\,. \tag{12}$$

In this way the angle ε acquires the same sign as the torsion τ. With easy substitution we then have

$$\sin\varepsilon = \frac{\boldsymbol{n}_1\wedge\boldsymbol{n}_2\times(P_2-P_1)}{\sigma\sin\zeta_1\sin\zeta_2}\,. \tag{13}$$

Carrying out the calculations we then have

$$\boldsymbol{n}_1\wedge\boldsymbol{n}_2 = s\left[\tau+\frac{s^2}{24}(\tau''-4\tau\omega^2)\right]\boldsymbol{t}+\frac{s^3}{8}(\varkappa'\tau-\tau'\varkappa)\boldsymbol{n}$$

$$+s\left[\varkappa+\frac{s^2}{24}(\varkappa''-4\varkappa\omega^2)\right]\boldsymbol{b} \tag{14}$$

and hence

$$n_1 \wedge n_2 \times (P_2 - P_1) = s^2 \tau - \frac{s^4}{24}(4\tau\omega^2 - \tau'') - \frac{s^6}{576}[\varkappa(\varkappa\tau'' - \tau\varkappa'') + 3\varkappa'(\varkappa\tau' - \varkappa'\tau)]$$

$$\frac{1}{\sigma \sin\zeta_1 \sin\zeta_2} = \frac{1}{s}\left[1 + \frac{7s^2\varkappa^2}{24} - \frac{s^4}{1152}(49\varkappa^2\tau^2 - 62\varkappa^4 - 7\varkappa'^2 - 24\varkappa\varkappa'')\right] \quad (15)$$

whence, finally, limiting ourselves to the terms written

$$\sin\varepsilon = s\left[\tau - \frac{s^2}{24}(4\tau^3 - 3\tau\varkappa^2 - \tau'')\right] \quad (16)$$

and

$$\varepsilon = s\left[\tau + \frac{s^2}{8}\left(\tau\varkappa^2 + \frac{\tau''}{3}\right)\right]. \quad (17)$$

5. Considering now the spherical triangle $C_1 n_2 n_1$, of which the angle at C_1 is equal to ε, and those at n_2 and n_1 are respectively η_2 and $180° - \eta_1$, we see that the sides $C_1 n_2$ and $C_1 n_1$ are respectively the zenith distance ζ_2 of P_1, and the supplement $180° - \zeta_1$ of the zenith distance ζ_1 of P_2, at P_2 and P_1 respectively. Hence, applying the analogies of Napier

$$\operatorname{tg}\frac{\eta_1 - \eta_2}{2} = -\operatorname{tg}\frac{\varepsilon}{2}\,\frac{\sin\dfrac{\zeta_1 - \zeta_2}{2}}{\sin\dfrac{\zeta_1 + \zeta_2}{2}}. \quad (18)$$

Making use of the expansion of arcsin, and noting that both ζ_1 and ζ_2 are rather close to 90°, we have from Eq. (8)

$$90° - \zeta_1 = \frac{s^2}{2\sigma}\left\{\varkappa - \frac{s}{6}\varkappa' - \frac{s^2}{12}\left[\varkappa\left(\frac{\varkappa^2}{2} + \tau^2\right) - \varkappa''\right]\right\}$$

$$90° - \zeta_2 = \frac{s^2}{2\sigma}\left\{\varkappa + \frac{s}{6}\varkappa' - \frac{s^2}{12}\left[\varkappa\left(\frac{\varkappa^2}{2} + \tau^2\right) - \varkappa''\right]\right\} \quad (19)$$

whence

$$\frac{\zeta_1 - \zeta_2}{2} = \frac{s^2}{12\sigma}\varkappa'\,; \qquad 90° - \frac{\zeta_1 + \zeta_2}{2} = \frac{s^2}{2\sigma}\left\{\varkappa - \frac{s^2}{12}\left[\varkappa\left(\frac{\varkappa^2}{2} + \tau^2\right) - \varkappa''\right]\right\} \quad (20)$$

and again

$$\frac{1}{\sin\dfrac{\zeta_1 + \zeta_2}{2}} = 1 + \frac{s^2\varkappa^2}{48}$$

$$\operatorname{tg}\frac{\varepsilon}{2} = \frac{s}{2}\left[\tau + \frac{s^2}{8}\left(\tau\varkappa^2 + \frac{2}{3}\tau^3 + \frac{\tau''}{3}\right)\right] \quad (21)$$

and thus

$$\eta_1 - \eta_2 = \frac{s^3 \varkappa' \tau}{12}\left[1 + \frac{s^2}{48}\left(9\varkappa^2 + 4\tau^2 - \frac{\varkappa'}{18} + 2\frac{\tau''}{\tau}\right)\right]. \tag{22}$$

The principal part of Dalby's term is thus given independently of the surface, except for a factor $s^3/12$, by the product of the torsion and the derivative of the first curvature of the surface in the direction considered.

6. In the case of the rotational ellipsoid we have immediately, neglecting terms in e^4 and higher,

$$\varkappa' = \frac{3e^2 \sin 2\varphi \cos\alpha}{2a}$$

$$\tau = \frac{e^2 \cos^2\varphi \sin 2a}{2a} \tag{23}$$

in which α is the azimuth of the geodesic at P_0, φ the latitude, e the eccentricity, and a the major axis of the meridional ellipse. In this case, the principal part of Dalby's term is written

$$\eta_1 - \eta_2 = \frac{s^3 e^4 \sin 2\varphi \cos^2\varphi \sin 2\alpha \cos\alpha}{16a^3}. \tag{24}$$

It is thus small to the 5th order in (s/a) and also in e^2. For example, for $\varphi = 45°$, $s = 1000$ km we would have

$$\eta_1 - \eta_2 = 0''.0008 .$$

Writing the formula in the following way

$$\eta_1 - \eta_2 = \frac{s^3 e^4 \sin 2\varphi \cos^2\varphi \sin\alpha \cos^2\alpha}{8a^3} \tag{24'}$$

and setting with a crude approximation

$$a\Delta\varphi = s\cos\alpha$$

$$a\cos\varphi\Delta\lambda = s\sin\alpha . \tag{25}$$

$\Delta\varphi$ and $\Delta\lambda$ being the difference of latitude and of longitude between P_1 and P_2, we arrive at Dalby's form

$$\eta_1 - \eta_2 = \frac{e^4(\Delta\varphi)^2 \Delta\lambda \sin 2\varphi \cos^3\varphi}{8}. \tag{24''}$$

Bibliography

Ivory J (1828) On the method in the trigonometrical survey for finding the difference of longitude of two stations very little different in latitude. Philos Mag, Ann Philos, December 1828:432

Tiarks Dr. (1828) On Mr Dalby's method of finding the difference of longitude between two points of a geodetical line on a spheroid, from the latitude of those points and the azimuths of the geodetical line at the same. Philos Mag, Ann Philos, November 1828:364

III Principles of Intrinsic Geodesy Applied to the Normal Reference Field

1 Principles of Intrinsic Geodesy Applied to the Field of Somigliana[1]

1. It is well known that the study of Geodesy is usually based on the use of particular mathematical surfaces of reference, ellipsoids, or spheroids, which are suggested by cosmological or mechanical theories, and are confirmed by experiment to be approximately the equilibrium figure of a particular level surface of the Earth's gravity field, the geoid.

In particular, the geometric approach to Geodesy, whose fundamental equations are those of triangulation, and of the measurement of astronomical latitude, longitude, and azimuth, has for centuries been dominated by the ellipsoidal concept. This was first conceived by Newton, and has basically taken the form of a flattened ellipsoid of rotation. On the other hand, the dynamic approach, whose fundamental operations are gravimetric and variometric measurements, was dominated up to 1930 by the concept which we may call spheroidal, due above all to the work of Bruns and of Helmert, and which took as its basis the mathematical surface of a spheroid of degree 14, but in fact little different from an ellipsoid, and whose justification rested purely on the convenience of halting the expansion of the potential, in powers of the radius vector, after the first few terms.

After 1930, following the decision taken by the International Association of Geodesy and Geophysics which met at Stockholm, the two approaches to Geodesy, geometric and dynamic, became fused together, in so far as from that time onward they were both referred to the same surface of comparison, the ellipsoid (and in particular the International Ellipsoid). This was done on the proposal of Cassinis, and was the culmination of the theoretical investigations of Pizzetti and Somigliana, investigations to which Silva and Cassinis gave concrete form for geodetic applications. With such an ellipsoid it is now possible to speak of a gravity vector, in virtue of the theorem of Stokes, and in resolving two distinct problems of Dirichlet (or of Stokes), with the determination of the potential field for which the ellipsoid is one of the equipotential surfaces. If this ellipsoid is the International Ellipsoid, the field takes the name of Normal Field.

With this result of capital importance from both the theoretical and the practical standpoint, and to which the contribution of Italian studies has been essential, it can be said that the need for a field of comparison in Geodesy has been completely defined, and leaves nothing to be desired.

2. Since, however, all the measurements made in operational Geodesy are referred not to this ideal field, but to the real field as it appears to observation,

[1] Originally published as: Marussi A (1950) Principi di Geodesia intrinseca applicati al campo di Somigliana. Boll Geodes Sci Affini no 2:1 – 8

the problem can arise as to whether it might not be convenient, and at times opportune, to face up to the study of the real gravity vector without having recourse to such an hypothetical comparison field.

This possibility undoubtedly exists from the theoretical aspect, for example by making use of vectorial and homographic methods, which not only do not need particular reference surfaces, but also avoid the use of any absolute system of Cartesian axes. This study, which is more to be reckoned as an exercise in Mathematical Physics than in Geodesy, can yet prove to be fruitless in practice when, in the end, vectors and homographies are removed from the possibility of direct measurement, and only scalar quantities are accessible to observation.

At this point we might think of proceeding to the measurement of the components of these absolute entities, vectors and homographies, with respect to appropriately chosen triads, in such a way that either these triads, or their relative components, should have an immediate meaning, compatible with the nature of the problem in hand, and suitable for carrying out the required experimental observations.

Having rejected a single Cartesian reference, as extraneous and inaccessible, we can, however, solve the problem of defining the triad if we have available in space a triple family of coordinate surfaces, which at every point determine the characteristic triads by the tangents to the respective lines of intersection.

In the case of the Earth's gravity field, the first family of coordinate surfaces which comes to mind is that of the equipotential surfaces, each surface being distinguished by the corresponding value of the potential W appropriate to it, and which will thus represent the first coordinate that we choose.

At each point on each equipotential surface it is possible to define, by astronomical observations, the direction of the normal to the surface, fixed by means of the usual angular coordinates φ and λ, latitude and longitude, referred to the Earth's astronomical axis. If we join into a surface all the points in space which have the same latitude, we shall have determined a φ surface; and similarly for λ.

Thus the φ and W surfaces intersect along curves which are called geodetic parallels; the λ and W surfaces intersect instead along curves called geodetic meridians; and the φ and λ surfaces along curves called isozeniths. The three congruences of curves so defined are the framework of the triads which are best adapted for the scalar representation of the respective vector and homographic components of the gravity field, when we wish to apply the experimental methods of Geodesy.

In particular, such triads are defined at each point by three vectors:

$$v_1 = \frac{\partial P}{\partial \varphi}; \qquad v_2 = \frac{\partial P}{\partial \lambda}; \qquad v_3 = \frac{\partial P}{\partial W}.$$

By projecting on these the specified absolute entities, the complexes of components take the name of tensors: simple tensors, if the entity projected is a vector, double tensors if this is a homography, and so on.

From the fact that the triads of vectors (which we shall call base vectors) are defined with the help of the triple family of surfaces (or congruences of lines) of which we have spoken, and which allow the triads to run, as it were, upon tracks from one point of space to another, then by using the relationships inhering in the absolute differential (or tensorial) calculus, it is not difficult to transport the properties of a tensor from one point to another of the gravity field. This allows the comparison of the same entity as it is manifest at different points, and hence all the consequent deductions which form the object of the study of Geodesy.

From an analytical viewpoint, the passage from one triad to the next infinitesimally close requires, in our case, the knowledge of 16 "coefficients of connexion" for which we usually adopt the so-called Christoffel symbols of the second kind. In the case of the most general coordinates this number would be 27, of which, however, only 18 differ from each other.

The study thus sketched in its broadest outlines can legitimately take the name of Intrinsic Geodesy, in as much as it has recourse only to entities which are directly measurable and which are inherently compatible with the object of study of Geodesy, rejecting any inaccessible reference.

3. Leaving aside now these considerations of the actual gravity field of the Earth, it could be interesting to apply the principles of Intrinsic Geodesy to the theoretical field of Somigliana, at the outset to provide a simple concrete example, and later also for the practical applications to which such principles may lead. A large part of the developments which follow can be applied not only to this theory, but in general to any field endowed with rotational symmetry, of which Somigliana's field is but a particular case.

We denote now by ρ, N, and g respectively the radius of curvature of the meridian, and of the prime vertical, and the gravity, and note that for fields of rotation about the same axis with respect to which the latitude and longitude are defined, the azimuths of the geodetic meridians and parallels are 0° and 90° respectively, and thus the meridians and parallels are at the same time the North and East lines. Setting for convenience

$$\varphi = y^1; \quad \lambda = y^2; \quad W = y^3$$

we shall have for the linear element ds

$$ds^2 = dP \times dP = \Sigma g_{rs} dy^r dy^s = \mathscr{E} d\varphi^2 + \mathscr{G} d\lambda^2 + 2\mathscr{L} d\varphi dW + \mathscr{N} dW^2$$

(the rectangular coefficients in $d\lambda$ are identically zero, from the rotational symmetry of the field), in which

$$\mathscr{E} = \rho^2; \quad \mathscr{G} = N^2 \cos^2\varphi; \quad \mathscr{L} = \frac{\rho}{g} \frac{\partial \lg g}{\partial \varphi};$$

$$\mathscr{N} = \frac{1}{g^2}\left\{1 + \left(\frac{\partial \lg g}{\partial \varphi}\right)^2\right\}; \quad \sqrt{\mathscr{E}\mathscr{G}} = \rho N \cos\varphi .$$

Considering the matrix $\| g_{rs} \|$ formed by the components of the fundamental metric tensor g_{rs}:

$$\| g_{rs} \| = \begin{Vmatrix} \mathscr{E} & 0 & \mathscr{L} \\ 0 & \mathscr{G} & 0 \\ \mathscr{L} & 0 & \mathscr{N} \end{Vmatrix}$$

the value of the corresponding determinant $\mathscr{D}^2$ is given by

$$\mathscr{D}^2 = |g_{rs}| = \left(\frac{\rho N \cos\varphi}{g}\right)^2 .$$

If we consider a particular equipotential surface $W = \text{const}$, the coefficients of its second fundamental form are

$$-dn \times dP = D d\varphi^2 + 2D' d\varphi d\lambda + D'' d\lambda^2$$

where $\boldsymbol{n}$ is the normal versor directed outwards, we have that

$$D = -\rho\,; \qquad D' = 0\,; \qquad D'' = -N\cos^2\varphi$$

and hence

$$\Delta = DD'' - D'^2 = DD'' = \rho N \cos^2\varphi\,.$$

Denoting by $(\boldsymbol{i}_1, \boldsymbol{i}_2, \boldsymbol{i}_3)$ the local astronomical versors ($\boldsymbol{i}_1$ and $\boldsymbol{i}_2$ are tangent to the equipotential surface considered, the first directed to the astronomical North, the second to the East; $\boldsymbol{i}_3 = \boldsymbol{n}$), the base vectors $\boldsymbol{v}_1 = \dfrac{\partial P}{\partial y^1}$ are written

$$\boldsymbol{v}_1 = \frac{\partial P}{\partial \varphi} = \rho \boldsymbol{i}_1\,; \qquad \boldsymbol{v}_2 = \frac{\partial P}{\partial \lambda} = N\cos\varphi\, \boldsymbol{i}_2\,; \qquad \boldsymbol{v}_3 = \frac{\partial P}{\partial W} = \frac{1}{g}\left(\frac{\partial \lg g}{d\varphi}\boldsymbol{i}_1 - \boldsymbol{i}_3\right).$$

The reciprocal vectors $\boldsymbol{v}^i$ defined by $\boldsymbol{v}^i \times \boldsymbol{v}_j = \delta^i_j$, in which δ^i_j represents Kronecker's symbol, are then given by

$$\boldsymbol{v}^1 = \frac{1}{\rho}\left(\boldsymbol{i}_1 + \frac{\partial \lg g}{\partial \varphi}\boldsymbol{i}_3\right); \qquad \boldsymbol{v}^2 = \frac{\boldsymbol{i}_2}{N\cos\varphi}\,; \qquad \boldsymbol{v}^3 = -g\boldsymbol{i}_3\,.$$

Further, the trigonometric tangent of the zenith distance of the isozenithal line is given by

$$\operatorname{tg} \nu = -\frac{\partial \lg g}{\partial \varphi}\,.$$

4. Given this, the Mainardi-Codazzi equations for the φ, λ, and W surfaces, which usually number 6, are reduced to 3 only, the other three being satisfied identically, and are written

$$\frac{\partial \rho}{\partial W} = -\frac{1}{g} - \frac{\partial^2}{\partial \varphi^2}\frac{1}{g}\,; \qquad \frac{\partial N}{\partial W} = -\frac{1}{g} + \operatorname{tg}\varphi \frac{\partial}{\partial \varphi}\frac{1}{g}\,; \qquad \frac{\partial N}{\partial \varphi} = (N - \rho)\operatorname{tg}\varphi$$

in which we must exclude the case where the poles are singular points, a case obviously of no geodetic interest.

The first two equations relate the increase of the principal radii of curvature with the dynamic height to the surface derivatives of gravity. If we have instead the orthometric height h, we can immediately apply the formulae given elsewhere and write

$$\frac{\partial \rho}{\partial h} = 1 + \frac{\partial \lg g}{\partial \varphi}\,\frac{\partial \lg \rho}{\partial \varphi} + g\frac{\partial^2}{\partial \varphi^2}\,\frac{1}{g}; \qquad \frac{\partial N}{\partial h} = 1 + \frac{N}{\rho}\operatorname{tg}\varphi \frac{\partial \lg g}{\partial \varphi}$$

since symbolically

$$\frac{\partial}{\partial h} = \frac{1}{\rho}\,\frac{\partial \log g}{\partial \varphi}\cdot\frac{\partial}{\partial \varphi} - g\frac{\partial}{\partial W}.$$

Turning now to the calculation of the Christoffel symbols of the second kind, we obtain immediately from their general expressions:

$$\Gamma^1_{11} = \frac{\partial}{\partial \varphi}\lg\frac{\rho}{g};$$

$$\Gamma^1_{22} = \frac{N}{\rho}\cos^2\varphi\frac{\partial}{\partial \varphi}\lg\frac{N}{g} + \sin\varphi\cos\varphi = \frac{N\cos^2\varphi}{\rho}\left(\operatorname{tg}\varphi - \frac{\partial \lg g}{\partial \varphi}\right)$$

$$\Gamma^1_{13} = \Gamma^1_{31} = \frac{\partial \lg \rho}{\partial W}; \qquad \Gamma^1_{33} = \frac{1}{\rho g}\,\frac{\partial^2 \lg g}{\partial \varphi\,\partial W}$$

$$\Gamma^2_{12} = \Gamma^2_{21} = -\frac{\rho}{N}\operatorname{tg}\varphi; \qquad \Gamma^2_{23} = \Gamma^2_{32} = \frac{\partial \lg N}{\partial W}$$

$$\Gamma^3_{11} = \rho g; \qquad \Gamma^3_{22} = gN\cos^2\varphi$$

$$\Gamma^3_{33} = -\frac{\partial \lg g}{\partial W} = \frac{-2\omega^2}{g^2} - \frac{1}{g}\left(\frac{1}{\rho} + \frac{1}{N}\right) - \frac{1}{\rho g}\left(\frac{\partial \lg g}{\partial \varphi}\right)^2$$

where the symbols not written are identically zero.

The surface Christoffel symbols of the second kind on a level surface, always referred to the usual intrinsic geographic coordinates, are then, as found in another form in a note of Tonolo (1939):

$$\bar{\Gamma}^1_{11} = \frac{\partial \lg \rho}{\partial \varphi}; \qquad \bar{\Gamma}^1_{22} = \frac{N}{\rho}\sin\varphi\cos\varphi$$

$$\bar{\Gamma}^2_{12} = \bar{\Gamma}^2_{21} = -\frac{\rho}{N}\operatorname{tg}\varphi$$

with all the others identically zero.

As a check, the following known relationships can be verified:

$$\Gamma^i_{1i} = \frac{\partial}{\partial \varphi}\lg \mathcal{D}; \qquad \Gamma^i_{2i} = \frac{\partial}{\partial \lambda}\lg \mathcal{D} = 0; \qquad \Gamma^i_{3i} = \frac{\partial}{\partial W}\lg \mathcal{D};$$

$$\bar{\Gamma}^\alpha_{1\alpha} = \frac{\partial}{\partial \varphi}\lg \bar{\mathcal{D}}; \qquad \bar{\Gamma}^\alpha_{2\alpha} = \frac{\partial}{\partial \lambda}\lg \bar{\mathcal{D}} = 0$$

(with summation for i from 1 to 3, and of α from 1 to 2; $\bar{\mathcal{D}} = \rho N\cos\varphi$).

In the same way, if we wish to calculate the second derivatives of the potential with respect to the orthogonal Cartesian coordinates x^1, x^2, x^3 related to the local astronomical triads, we have easily the relationships $\left(\text{in which } W_{rs} = \dfrac{\partial^2 W}{\partial x^r\,\partial x^s}\right)$:

$$W_{11} = -\frac{g}{\rho} \qquad W_{12} = 0 \qquad W_{22} = -\frac{g}{N}$$

$$W_{23} = 0 \qquad W_{13} = -\frac{g}{\rho}\frac{\partial \lg g}{\partial \varphi} = -\frac{g}{R}$$

$$W_{33} = 2\omega^2 + g\left(\frac{1}{\rho} + \frac{1}{N}\right) = -\frac{g}{\rho}\left(\frac{\partial \lg g}{\partial \varphi}\right)^2 + g^2\frac{\partial \lg g}{\partial W}$$

where R is the radius of curvature of the line of force

$$\frac{1}{R} = \frac{1}{\rho}\frac{\partial \lg g}{\partial \varphi} = -\frac{1}{\rho}\operatorname{tg} v\,.$$

The quantities indicated in the preceding chapters by h, k, H_1 and H_2 then have the following values

$$h = \frac{N\cos\varphi}{g}\frac{\partial \lg g}{\partial \varphi} \qquad k = 0$$

$$H_1 = -\frac{1}{g}\frac{\partial \lg g}{\partial \varphi} \qquad H_2 = 0\,.$$

For the covariant components of the gravimetric gradient we find immediately

$$\mathscr{G}_1 = \frac{\partial g}{\partial \varphi} = g\frac{\partial \lg g}{\partial \varphi} = -g\operatorname{tg} v = -\frac{\rho g}{R}$$

$$\mathscr{G}_2 = 0 \qquad \mathscr{G}_3 = \frac{\partial g}{\partial W} = g\frac{\partial \lg g}{\partial W} = -g\Gamma_{33}^3\,.$$

Finally, the covariant components w_{rs} of the spatial derivative of the gravity vector are given, as is well known, by

$$w_{rs} = -\Gamma_{rs}^3\,.$$

5. Applying all this to Somigliana's theory, we shall limit ourselves to finding surface values on the base ellipsoid of all the entities considered up to now. It will suffice to substitute in the equations we have found the following expressions

$$\rho = \frac{a(1-e^2)}{(1-e^2\sin^2\varphi)^{3/2}} \qquad N = \frac{a}{(1-e^2\sin^2\varphi)^{1/2}}$$

$$g = g_0(1 + \alpha\sin^2\varphi - \beta\sin^2 2\varphi)$$

in order to obtain

$$\frac{\partial \rho}{\partial \varphi} = \frac{3}{2}\frac{ae^2(1-e^2)\sin 2\varphi}{(1-e^2\sin^2\varphi)^{5/2}} = \frac{3}{2}\rho\frac{e^2\sin 2\varphi}{1-e^2\sin^2\varphi}$$

$$\frac{\partial \lg \rho}{\partial \varphi} = \frac{3}{2} \frac{e^2 \sin 2\varphi}{1 - e^2 \sin^2 \varphi}$$

$$\frac{\partial N}{\partial \varphi} = (N - \rho) \operatorname{tg} \varphi = \frac{a e^2 \sin 2\varphi}{2(1 - e^2 \sin^2 \varphi)^{3/2}} = N \frac{e^2 \sin 2\varphi}{2(1 - e^2 \sin^2 \varphi)}$$

$$\frac{\partial \lg N}{\partial \varphi} = \frac{e^2 \sin 2\varphi}{2(1 - e^2 \sin^2 \varphi)}$$

$$\frac{1}{g} = \frac{1}{g_0} \left\{ 1 + (\alpha^2 - \alpha) \sin^2 \varphi + \left(\beta - \frac{\alpha^2}{4} \right) \sin^2 2\varphi \right\}$$

$$= \frac{1}{g_0} \left\{ 1 + \frac{3\alpha^2}{8} + \frac{\beta - \alpha}{2} + \frac{\alpha - \alpha^2}{2} \cos 2\varphi + \frac{1}{8} (\alpha^2 - 4\beta) \cos 4\varphi \right\}$$

$$\frac{\partial g}{\partial \varphi} = g_0 \sin 2\varphi (\alpha - 4\beta \cos 2\varphi)$$

$$\frac{\partial \lg g}{\partial \varphi} = \left(\alpha - \frac{\alpha^2}{2} \right) \sin 2\varphi + 2 \left(\frac{\alpha^2}{8} - \beta \right) \sin 4\varphi$$

$$\frac{\partial^2 \log g}{\partial \varphi^2} = (2\alpha - \alpha^2) \cos 2\varphi + (\alpha^2 - 8\beta) \cos 4\varphi$$

$$\frac{\partial}{\partial \varphi} \frac{1}{g} = \frac{\alpha^2 - \alpha}{g_0} \sin 2\varphi + \frac{1}{2 g_0} (4\beta - \alpha^2) \sin 4\varphi$$

$$\frac{\partial^2}{\partial \varphi^2} \frac{1}{g} = \frac{2}{g_0} \left\{ (\alpha^2 - \alpha) \cos 2\varphi + (4\beta - \alpha^2) \cos 4\varphi \right\}$$

where in the various expressions that are functions of g we have kept only the terms in α, β and α^2.

If we are to refer to the "normal field", which as we know is based on the International Ellipsoid and the appropriate value of equatorial gravity determined by Silva, we need only substitute in all the foregoing formulae the following numerical values:

$a = 6{,}378{,}388$ m $\qquad e^2 = 6.72267 \times 10^{-3}$

$g_0 = 978.049$ Gal $\qquad \alpha = 5.28840 \times 10^{-3}$

$\beta = 5.9 \times 10^{-6}$

$\omega = 72.9213 \times 10^{-6}\,\mathrm{s}^{-1} \qquad 2\omega^2 = 10.6349 \times 10^9\,\mathrm{s}^{-2}$

$\qquad = 106{,}349\, E\,.$

Of the three equations of Mainardi-Codazzi which, as has been said, express the conditions of integrability of the field, one is satisfied identically, and the other two assume the following form, as has already been shown elsewhere:

$$\frac{\partial \rho}{\partial W} = -\frac{1}{g_0}\left\{1+\frac{3\alpha^2}{8}+\frac{\beta-\alpha}{2}+\frac{3}{2}(\alpha^2-\alpha)\cos 2\varphi+\frac{15}{2}\left(\beta-\frac{\alpha^2}{4}\right)\cos 4\varphi\right\}$$

$$= -1.0197538\times 10^{-3}+8.0677\times 10^{-6}\cos 2\varphi+8.4\times 10^{-9}\cos 4\varphi(\mathrm{Gal}^{-1})$$

$$\frac{\partial N}{\partial W} = -\frac{1}{g_0}\left\{1-\frac{3}{4}\alpha^2+\frac{\alpha}{2}+\beta+\frac{3\alpha^2-\alpha+8\beta}{2}\cos 2\varphi+3\left(\beta-\frac{\alpha^2}{4}\right)\cos^2 2\varphi\right\}$$

$$= -1.0251317\times 10^{-3}+2.6876\times 10^{-6}\cos 2\varphi$$

$$+3.4\times 10^{-9}\cos^2 2\varphi(\mathrm{Gal}^{-1})\,.$$

As a check, we should have at the poles ($\varphi = \pm 90°$), $\frac{\partial \rho}{\partial W}=\frac{\partial N}{\partial W}$.

These latter formulae allow us to calculate to a first approximation the radii of curvature of an equipotential surface close to that of the base ellipsoid.

Bibliography

Bruns H (1878) Die Figur der Erde. Veröff d Preuß Geod Inst, Berlin

Helmert FR (1880) Die mathematischen und physikalischen Theorien der Höheren Geodäsie – Einleitung und I. Teil: Die mathematischen Theorien. B. G. Teubner, Leipzig

Helmert FR (1884) Die mathematischen und physikalischen Theorien der Höheren Geodäsie – II. Teil: Die Physikalischen Theorien, mit Untersuchungen über die mathematische Erdgestalt auf Grund der Beobachtungen. B. G. Teubner, Leipzig

Marussi A (1949) Fondements de géométrie différentielle absolue du champ potentiel terrestre. Bull Géodés Assoc Int Géodésie no 14

Marussi A (1950) Sulla variazione con l'altezza dei raggi di curvatura principali nella teoria di Somigliana. Boll Geodes Ist Geogr Milit, Florence, no 1

Tonolo A (1939) Trasporto delle coordinate geografiche e dell'azimut lungo un arco di linea qualunque di un ellissoide di rotazione. Rend R Accad Naz Lincei 29: Ser 6, 1st sem, fasc II, June 1939

2 The First Fundamental Problem of Geodesy Developed for the Field of Somigliana[1]

1. What is known as the first fundamental problem of Geodesy comprises the determination of the geographic coordinates φ and λ (latitude and longitude) of the end P of a geodesic of finite length s drawn out on a surface Σ, and its azimuth at P, where the geographic coordinates φ_0 and λ_0 at its other end P_0 are known, as well as the azimuth α_0 at this latter point.

Such a problem has for some time been resolved when the surface is that of a rotational ellipsoid, which is the case of most interest for practical applications in operational Geodesy. We owe to Legendre the series expansions which are still today used almost exclusively and which he published in the Memoirs of the Academy of Paris in 1806. These expansions thus go under the name of Legendre expansions (or as also of Delambre's formulae). From that time on, many Geodesists have concerned themselves with this fundamental problem, with the object of making the formulae ever more rigorous, or more apt for practical applications over sufficiently wide regions; and among these we recall the names of Delambre, Bessel, Andrae, Schreiber, Helmert, Börsch, Jadanza, Grabowski, Krüger.

Recently, Legendre expansions have been taken up again from a more general point of view by Tonolo (1939), who has developed them for any given curve (rather than just for a geodesic) drawn out on the rotational ellipsoid, the curve being defined not only by one extremity, the tangent there, and the length, but in addition by the tangential curvature which is considered to be known as a function of the arc. The late lamented Prof. A. Marcantoni has considered the expansions with reference to a geodesic traced out on a triaxial ellipsoid (1948); and a further extension to the case of any curve whatever traced out on any surface whatever is given in Chapter I.3, this Volume.

2. The consideration of the transport of the geographic coordinates and the azimuth in the latter and most general case (which can immediately be reduced to that of a geodesic by setting the tangential curvature to zero) can be of interest on all occasions when it is desired to refer geodetic research to a gravity equipotential surface, rather than to the conventional ellipsoidal surface. In this case, too, as in the classical case of the ellipsoid, the transport can occur only when it is possible to speak of a curve (a geodesic or not) drawn *on* the surface, and whose length s is measured along the curve itself, always lying on that surface.

---

[1] Originally published as: Marussi A (1950) Il primo problema fondamentale della Geodesia ampliato nel campo di Somigliana. Geofis Pura Appl 18:3–11 (dedicated to Prof Carlo Somigliana)

This condition can be fulfilled only rarely in practice; it is fulfilled in particular with sufficient rigor in the direct measurement of geodetic bases laid out along extensive plains; but it is vitiated as soon as we consider the sides of a triangulation, for whose measurement (necessarily indirect) we must carry out operations which, in general, oblige us to leave the base equipotential surface, and to involve a multiplicity of level surfaces.

It is well known that this conceptual difficulty is overcome in operational Geodesy by means of a series of reductions (or corrections) which are applied to the quantities observed either directly or indirectly. All of these are based, however, on simplifying hypotheses, which do not in general adhere strictly to reality, and which thus undermine the rigor of the conclusions.

To these considerations we can add another: recent progress in Physics has permitted the employment of methods and of instruments which, already today, allow us to measure, with relative ease and adequate precision, distances in space (and no longer on surfaces as in classical Geodesy) along curves that approximate – with very minor corrections – to straight lines, and thus to space geodesics. As is well known, these methods are based on very short waves of the Radar type, or on the interference of modulated light waves, as in the Swedish apparatus of Bergstrand, called a geodimeter.

3. The reasons we have briefly recalled prompt us to examine the convenience of dealing with geodetic problems in a three-dimensional field, a field which is, after all, their natural habitat. Only thus can we avoid that series of either concealed or apparent approximations, and unsatisfactory reductions, to which we must resort if we wish to make use of a single surface of comparison.

This approach has been followed in several of my works cited in the bibliography. In these we have given in tensor form, that most apt for these researches of a general character, the fundamental equations valid for any potential field, and thus valid also for the actual gravity field of the Earth. We shall have occasion to return to these equations in what follows.

4. In classical Geodesy, then, we are faced with the need of making use of a conventional surface of comparison, the ellipsoid, as the site of computations to a first approximation, while at the same time we require a simple model to which to refer the "departures" or "anomalies" of the real field. Following the approach indicated, it seems opportune to make use of an analogous three-dimensional model, to be employed equally for computations to a first approximation, as well as a comparison for the actual gravity field of the Earth.

Today, fortunately, there can no longer be any doubt as to the choice of such a model, which must necessarily fall on the field of Somigliana, particularly if we wish to ensure coherence with the researches of classical Geodesy, both geometric and dynamic.

Geometric Geodesy has for centuries made use of the rotational ellipsoid; in 1924, at the Madrid Assembly, the International Association of Geodesy and Geophysics decided on the universal adoption for this ellipsoid of the parameters determined by Hayford on the basis of isostatic theories. To this ellipsoid, known as international, we today refer the greater part of the measurements and computations being undertaken in geometric Geodesy throughout the world.

In addition, from 1930 the international ellipsoid has also become the fundamental surface for the research of dynamic Geodesy, ever since Somigliana, crowning the preceding studies of Pizzetti, completely solved the problem of determining a gravitational field superimposed on a field of centrifugal forces, and of which one of the equipotential surfaces is that of an ellipsoid. The uniqueness of such a field is assured, but only under these conditions, by Stokes' theorem as generalized by Poincaré. From this fundamental research there arose, for the first time, the possibility of unifying the geometric and the dynamic approaches in Geodesy, which up till then had followed two divergent paths, the first depending on the ellipsoid, and the second on the spheroids of Helmert and of Bruns.

Somigliana and Lambert pointed out almost simultaneously the chance of making use of this result of capital importance in operational Geodesy; but it was on the basis of a concrete proposal of Cassinis (1930) that the International Association of Geodesy and Geophysics, at its Assembly in Stockholm in 1930, decided to adopt a formula for the gravity on the ellipsoid that was derived from the truncation at the first terms of the series expansion of Somigliana's rigorous formula. The parameters of this formula had been discussed and determined by Silva, being founded on the international ellipsoid. With this formula, and with these parameters, Cassinis calculated the Tables which present the so-called normal gravity, now universally adopted.

Professor Volta has observed that it might have been preferable to have given up the approximate formula which yields the value of gravity, and to have assumed instead the greater difficulties of the numerical calculation implicit in Somigliana's exact formula. To this end, he calculated the values of gravity which would have resulted with such an exact formula[2]. Such an observation takes on added importance when, if we wish to extend the formulas of Geodesy to the third dimension, we must consider (as we shall see) the successive derivatives of the modulus of gravity with respect to latitude. Here, doubt can arise as to whether this calculation might not yield significant differences were the approximate formula used rather than the exact one.

5. For the effective application of the concepts which have been briefly outlined here, it is opportune to make use of the methods of differential geometry – as does the classical Geodesy of surfaces. Even better is to employ the tensor calculus, which brings greater clarity and simplicity to the formulae.

In this respect, there have been developed, in the various works listed in the bibliography, the computations and the fundamental formulae, taken with reference to any potential field, which can therefore be identified with the real gravity field as it appears to observation. Several formulae are specified for the

---

[2] The formulae to which we have referred are the following:

(Somigliana's exact formula) $$g = \frac{a g_e \cos^2\varphi + b g_p \sin^2\varphi}{\sqrt{a^2\cos^2\varphi + b^2\sin^2\varphi}}$$

(approximate formula) $$g = g_e(1 + \eta_1 \sin^2\varphi - \tfrac{1}{4}\eta_2 \sin^2 2\varphi)$$

in which a and b are the semi-axes (equatorial or polar) of the ellipsoid, g_e and g_p the equatorial and polar gravity, η_1 and η_2 are two parameters whose expressions will be given in the following particular case in which the field is endowed with rotational symmetry, of application thus also to Somigliana's field. In all these researches, reference is made to the coordinates most familiar to geodesists, the latitude φ, the longitude λ, and the potential W, defined in an intrinsic manner, and thus without recourse to absolute reference axes that are inaccessible.

A first coordinate family is thus made up of the level surfaces themselves, each one being distinguished by the value of the potential W that is proper to it; a second family ($\varphi = \text{const}$) comprises the surfaces of the loci of points at which the vertical $\boldsymbol{n}$ forms a fixed angle ($90° - \varphi$) with the Earth's axis, with an analogous definition for the third family ($\lambda = \text{const}$). Regarding the potential W, we observe again that the difference between the values W_1 and W_2, relative to two points P_1 and P_2, measures the so-called *dynamic dislevelment,* except for a conventional factor, constant for the whole Earth. This dynamic dislevelment results from precision geometric levelling, after the directly observed values have had added the so-called dynamic correction. Denoting by Q_1 and Q_2 the dynamic heights of P_1 and P_2 (which are conventionally expressed in metres), we set

$$Q_2 - Q_1 = -\frac{1}{g_{0,45}}(W_2 - W_1)\,, \tag{1}$$

in which $g_{0,45}$ is the normal gravity on the ellipsoid at latitude 45° ($g_{0,45} = 980{,}6294$ Gal from the Tables of Cassinis).

6. Here we shall take up again these procedures, this time referring them to a rotational field, and applying them to the first fundamental problem of Geodesy developed in space for a geodesic (or straight line). Application to Somigliana's field follows immediately, it being only necessary to specify on the *base surface* the principal radii of curvature as those of the ellipsoid, and the modulus of gravity which results from Somigliana's formula.

This problem can be enunciated as follows: To determine the geographic coordinates φ and λ (latitude and longitude) and the potential W (or the dynamic height Q) of one end P of a straight line (space geodesic) of finite length s, the azimuth α of its vertical plane and the zenith distance at P, if the corresponding elements are known at the other end P_0.

7. The general solution of this problem has already been given for any field whatever in Chapter I.4, this Volume, and under yet more general conditions for the curve of transport in Chapter I.5, this Volume; it will suffice here to specify the meanings of the various symbols which are involved in the equations of the solution.

Denoting respectively by y^1, y^2, y^3 the latitude φ, the longitude λ, and the potential W, and setting $\lambda^i = dy^i/ds$, the expansions previously given can be summarized, limited to the first few terms, in the following:

$$y^r = y_0^r + s\lambda_0^r - \frac{s^2}{2!}\Gamma_{ij}^r \lambda_0^i \lambda_0^j + \dots \tag{2'}$$

$$\lambda^r = \lambda_0^r - s\Gamma_{ij}^r \lambda_0^i \lambda_0^j + \dots \tag{2''}$$

in which Γ_{ij}^r represents Christoffel's symbol of the second kind computed at P_0, and where it is to be understood that the italic indices can take values from 1 to 3, and further, that summation takes place with respect to all the (dummy) indices which appear twice in the same expression, once above and once below.

In the case of a rotational field, denoting by ρ and N respectively the radii of curvature of the meridian and of the prime vertical, we have (Chap. III.1, this Vol.) for the vectors of the mobile triad linked to the coordinates y^1, $\boldsymbol{v}_i = \partial P / \partial y^i$:

$$\boldsymbol{v}_1 = \rho \boldsymbol{i}_1 ; \qquad \boldsymbol{v}_2 = N \cos\varphi \boldsymbol{i}_2 ; \qquad \boldsymbol{v}_3 = \frac{1}{g}\left(\frac{\partial \lg g}{\partial \varphi} \boldsymbol{i}_1 - \boldsymbol{i}_3\right) \tag{3}$$

in which $(\boldsymbol{i}_1, \boldsymbol{i}_2, \boldsymbol{i}_3)$ represent the local astronomical triad of orthogonal versors ($\boldsymbol{i}_3$ directed toward the zenith, $\boldsymbol{i}_1$ to the North, and $\boldsymbol{i}_2$ to the East). For the reciprocal vectors $\boldsymbol{v}^j$ defined by

$$\boldsymbol{v}_i \times \boldsymbol{v}^j = \delta_i^j \qquad \text{(Kronecker's symbol)} \tag{4}$$

$$\boldsymbol{v}^1 = \frac{1}{\rho}\left(\boldsymbol{i}_1 + \frac{\partial \lg g}{\partial \varphi} \boldsymbol{i}_3\right) ; \qquad \boldsymbol{v}^2 = \frac{\boldsymbol{i}_2}{N \cos\varphi} ; \qquad \boldsymbol{v}^3 = -g \boldsymbol{i}_3 . \tag{5}$$

From this, the fundamental metric tensor is written (omitting the components that are identically zero)

$$g_{11} = \rho^2 ; \qquad g_{22} = N^2 \cos^2\varphi ; \qquad g_{33} = \frac{1}{g^2}\left\{1 + \left(\frac{\partial \lg g}{\partial \varphi}\right)^2\right\} ;$$

$$g_{13} = \frac{\rho}{g} \frac{\partial \lg g}{\partial \varphi} \tag{6}$$

and the Christoffel symbols take the form

$$\Gamma_{11}^1 = \frac{\partial}{\partial \varphi} \lg \frac{\rho}{g} ; \qquad \Gamma_{22}^1 = \frac{N \cos^2\varphi}{\rho}\left(\operatorname{tg}\varphi - \frac{\partial \lg g}{\partial \varphi}\right) ; \qquad \Gamma_{13}^1 = \frac{\partial \lg \rho}{\partial W} ;$$

$$\Gamma_{33}^1 = \frac{1}{g\rho} \frac{\partial^2 \lg g}{\partial \varphi \, \partial W} ; \qquad \Gamma_{12}^2 = -\frac{\rho}{N} \operatorname{tg}\varphi ; \qquad \Gamma_{23}^2 = \frac{\partial \lg N}{\partial W} ;$$

$$\Gamma_{11}^3 = g\rho ; \qquad \Gamma_{22}^3 = g N \cos^2\varphi ; \qquad \Gamma_{33}^3 = -\frac{\partial \lg g}{\partial W} ; \tag{7}$$

where those not written are identically zero.

The conditions of integrability then lead to the fundamental equations

$$\frac{\partial \rho}{\partial W} = -\frac{1}{g} - \frac{\partial}{\partial \varphi^2} \frac{1}{g} ; \qquad \frac{\partial N}{\partial W} = -\frac{1}{g} + \operatorname{tg}\varphi \frac{\partial}{\partial \varphi} \frac{1}{g} ;$$

$$\frac{\partial N}{\partial \varphi} = (N - \rho) \operatorname{tg}\varphi . \tag{8}$$

The first two permit the calculation of all the derivatives of the geometric elements of the field with respect to the potential (or to the dynamic height); and to these we can add the mechanical condition

$$\frac{\partial \lg g}{W} = \frac{2\omega^2}{g^2} + \frac{1}{g}\left(\frac{1}{\rho} + \frac{1}{N}\right) + \frac{1}{g\rho}\left(\frac{\partial \lg g}{\partial \varphi}\right)^2 \tag{9}$$

which arises from the equation $\operatorname{div} \boldsymbol{g} = 2\omega^2$ (*in vacuo*), and in fact expresses the Bruns equation (where ω is the angular velocity of rotation of the Earth $= 72.913 \times 10^{-6}\,\mathrm{s}^{-1}$).

This relationship permits, in turn, the calculation of the derivatives of the modulus of gravity with respect to the potential (or to the dynamic height).

8. As for the λ^i, putting $\boldsymbol{t} = \dfrac{P - P_0}{s}$, and denoting by α_0 the azimuth of the vertical plane containing $\boldsymbol{t}$ at P_0, and by z_0 the zenith distance of $\boldsymbol{t}$, we have

$$\boldsymbol{t} = \boldsymbol{i}_1 \sin z_0 \cos \alpha_0 + \boldsymbol{i}_2 \sin z_0 \sin \alpha_0 + \boldsymbol{i}_3 \cos z_0 \tag{10}$$

in which it is evident that the astronomic versors also refer to the point P_0. Then, recalling that $\lambda^i = \boldsymbol{t} \times \boldsymbol{v}^i$ we have immediately

$$\left.\begin{aligned} &\lambda^1 = \frac{1}{\rho}\left(\sin z \cos \alpha + \cos z \frac{\partial \lg g}{\partial \varphi}\right); \\ &\lambda^2 = \frac{\sin z \sin \alpha}{N \cos \varphi}; \qquad \lambda^3 = -g \cos z; \end{aligned}\right\} \tag{11}$$

in which the indices are taken relative to P_0.

9. Substitution of the values found in the general expansions of Legendre given above, leads to the following final formulae, in which for simplicity we write

$$s\lambda^1 = p; \qquad s\lambda^2 = q; \qquad s\lambda^3 = r: \tag{12}$$

$$\left.\begin{aligned} \varphi = {} & \varphi_0 + p - pr\frac{\partial \lg \rho}{\partial W} - \frac{p^2}{2}\frac{\partial}{\partial \varphi}\lg\frac{\rho}{g} - \frac{q^2 N \cos^2 \varphi}{2\rho}\left(\operatorname{tg}\varphi - \frac{\partial \lg g}{\partial \varphi}\right) \\ & - \frac{r^2}{2g\rho}\frac{\partial^2 \lg g}{\partial \varphi \partial W} + \dots \\ \lambda = {} & \lambda_0 + q\left(1 + \frac{p\rho}{2N}\operatorname{tg}\varphi - \frac{q}{2}\frac{\partial \lg N}{\partial W}\right) + \dots \\ W = {} & W_0 + r - \frac{1}{2}\left(p^2 g\rho + q^2 g N \cos^2 \varphi - r^2 \frac{\partial \lg g}{\partial W}\right) + \dots \end{aligned}\right\}. \tag{13}$$

Quite analogous formulae hold for the expansions of λ^i.

As an example, we shall focus attention on the last equation; taking note of (1), and setting

$$\gamma = \frac{g}{g_{0.45}} \tag{14}$$

it can also be written as

$$Q = Q_0 + s\gamma_0 \cos z_0 + \frac{s^2\gamma_0}{2}\left\{\frac{\sin^2 z_0}{R_{\alpha 0}} - \left(\frac{\cos^2 z_0}{R_{m0}} + \frac{2\omega^2}{g_0}\right) + \frac{\sin 2z_0 \cos\alpha_0}{R_0}\right\} + \ldots \tag{15}$$

in which R_α is the radius of curvature of the normal section of azimuth α of the equipotential surface. R_m is the mean radius of curvature:

$$\frac{1}{R_m} = \frac{1}{\rho} + \frac{1}{N} \tag{16}$$

and R is the radius of curvature of the line of force:

$$\frac{1}{R} = \frac{1}{\rho}\,\frac{\partial \lg g}{\partial \varphi}. \tag{17}$$

As is evident, the first term within the parenthesis is that which corresponds to the curvature correction of the usual formula for the trigonometric transport of height; the second, by contrast, takes account of the variation of gravity with height, and the third of the effect due to the curvature of the lines of force.

10. For the reciprocal zenith distance we would have

$$\cos z = \frac{g_0}{g}\cos z_0 + s\frac{g_0}{g}\left(\frac{\sin^2 z_0}{R_{\alpha 0}} + \frac{\sin 2z_0 \cos\alpha_0}{R_0} - \frac{\cos^2 z_0}{R_{m0}} - \frac{2\omega^2}{g_0}\right) + \ldots; \tag{18}$$

and its computation requires, as is seen, a knowledge of g at P. This can be had from the expansion

$$g = g_0 + s\frac{\partial g}{\partial y^i}\lambda^i + \frac{s^2}{2!}\left(\frac{\partial^2 g}{\partial y^i \partial y^j} - \frac{\partial g}{\partial y^r}\Gamma^r_{ij}\right)\lambda^i\lambda^j + \ldots \tag{19}$$

in which, as usual, all the coefficients of the powers of s are calculated at P_0.

In the same way, the reciprocal azimuth can be found, having once found the λ^i at P from the series expansions, from

$$\operatorname{tg}\alpha = \frac{\lambda^2 N\cos\varphi}{\rho\lambda^1 + \dfrac{\lambda^3}{g}\,\dfrac{\partial}{\partial\varphi}\lg g}, \tag{20}$$

where here the values of the coefficients are now referred to the point P; and their computation can be effected with formulas strictly analogous to that given for the transport of g.

It is also clear that the radii of curvature at P_0, as also the value of g at the same point, will need to be calculated – if P_0 does not lie on the base surface – by ordinary series expansions departing from this surface, on which the derivatives $\dfrac{\partial\rho}{\partial y^i}, \dfrac{\partial N}{\partial y^i}, \dfrac{\partial g}{\partial y^j}, \dfrac{\partial^2\rho}{\partial y^i\partial y^j}$, etc, have been found from the fundamental formulae already referred to.

Hence it is seen how all the calculations that have been mentioned can be reduced to those of the derivatives with respect to latitude of the radii of curvature and of the gravity *on the surface taken as fundamental*; since all the other derivatives which are involved can be expressed as functions of these.

Application to Somigliana's field is thus made very easy, and comes down simply to specifying ρ, N, and g on the base ellipsoid according to the well-known formulae which express these quantities as a function of latitude. These formulae are summarized here:

$$\left.\begin{aligned} &\rho=\frac{a(1-e^2)}{(1-e^2\sin^2\varphi)^{3/2}}\,; \qquad N=\frac{a}{(1-e^2\sin^2\varphi)^{1/2}}\,; \\ &g=\frac{a\,g_e\cos^2\varphi+b\,g_p\sin^2\varphi}{\sqrt{a^2\cos^2\varphi+b^2\sin^2\varphi}}\cong g_e\left(1+\eta_1\sin^2\varphi-\frac{1}{4}\eta_2\sin^2 2\varphi\right) \end{aligned}\right\} \tag{21}$$

where $e^2=(a^2-b^2)/a^2$; and then

$$\left.\begin{aligned} &g_p=\frac{b}{a}g_e+\frac{5}{2}b\omega^2\left(1+\frac{9}{35}e^2+\frac{47}{180}e^4+\dots\right); \\ &\eta_1=\frac{e^2}{2}(1+\varepsilon)+\frac{3}{8}e^4; \\ &\frac{1}{4}\eta_2=\frac{3}{32}e^4+\frac{1}{8}\varepsilon e^2; \end{aligned}\right\} \tag{22}$$

with

$$\varepsilon=\frac{b}{a}\frac{g_p}{g_e}-1\,. \tag{23}$$

If, in particular, the values chosen for a and e^2 are those for the international ellipsoid, and for g the equatorial value determined by Silva, and accepted internationally, then the field can be styled as *normal*. The relevant values are:

$$\left.\begin{aligned} &a=6378388\ \text{m}\,; \qquad e^2=6.72267\times10^{-3} \\ &g_e=978.049\ \text{Gal} \end{aligned}\right\}. \tag{24}$$

Bibliography

Cassinis G (1930) Sur l'adoption d'une formule internationale pour la pesanteur normale. Bull Géodés no 26:40

Marcantoni A (1948) Sul trasporto delle coordinate geografiche e degli azimut sopra un ellissoide a tre assi. Boll Geodes Ist Geogr Milit, Florence, no 2

Marussi A (1949) Fondements de géométrie différentielle absolue du champ potentiel terrestre. Bull Géodés no 14

Marussi A (1950) Sulla variazione con l'altezza dei raggi di curvatura principali nella teoria di Somigliana. Boll Geodes Ist Geogr Milit, Florence, no 1

Silva G (1928) Sulla definizione della gravità normale. Rend R Accad Naz Lincei 8: Ser 6, 2nd sem, fasc 12

Silva G (1930) Sulla formula della gravità normale. Rend R Accad Naz Lincei 11: Ser 6, 1st sem, fasc 7

Somigliana C (1929) Teoria del campo gravitazionale dell'ellissoide di rotazione. Mem Soc Astron Ital 4

Somigliana C (1946) Sviluppi in serie delle espressioni della gravità e determinazione gravimetrica delle costanti del geoide. Mem Soc Astron Ital 18: no 1

Somigliana C (1948) Costanti gravimetriche e misure di gravità. Nuovo Cimento 5: no 3

Tonolo A (1939) Trasporto delle coordinate geografiche e dell'azimut lungo un arco di linea qualunque di un ellissoide di rotazione. Rend R Accad Naz Lincei 29: Ser 6, 1st sem, fasc 11

3 Differential Parameters in the Normal Gravity Field[1]

1. Given a system of general coordinates x^i $(i = 1, 2, 3)$ in the three-dimensional metric space, at each point P two triads of reference vectors are defined:

$$\text{the base triad} \qquad v_i = \frac{\partial P}{\partial x^i} \tag{1}$$

$$\text{the reciprocal triad} \qquad v^i = \operatorname{grad} x^i . \tag{2}$$

The two triads are connected by the following equations:

$$v_i \cdot v_j = g_{ij}; \qquad v_i \cdot v^j = \delta_i^j; \qquad v^i \cdot v^j = g^{ij} \,{}^{2} \tag{3}$$

in which g_{ij} is the metric tensor, g^{ij} its reciprocal, and δ_i^j the Kronecker's symbol.

For an elementary *displacement vector* we have

$$dP = \frac{\partial P}{\partial x^i} dx^i = v_i dx^i \tag{4}$$

and for a *gradient vector*

$$\operatorname{grad} \varphi = \frac{\partial \varphi}{\partial x^i} \operatorname{grad} x^i = v^i \frac{\partial \varphi}{\partial x^i} . \tag{5}$$

The natural function of the base vectors is therefore to describe displacement vectors or contravariant tensors; and the natural function of the reciprocal vectors is to describe gradient vectors or covariant tensors.

The distinction between the two kinds of vectors and tensors is essential in Affine Geometry; but it still preserves a definite character in Metric Geometry.

2. As has been shown in a paper dedicated to prof. Tauno Kukkamäki on the occasion of this 70th birthday, the two reference triads are indeed not equivalent since:

a) the base vectors v_i are not in general gradient vectors, as the reciprocal vectors are;
b) the reciprocal vectors v^i are not in general the vectors of the principal triad of any triple family of surfaces, as the base vectors are.

[1] Originally published as: Marussi A (1980) Differential parameters in the normal gravity field. Rend Accad Naz Lincei 69: Ser 8, fasc 6, 410 – 414

[2] Summation over duplicated indices throughout the paper; the scalar product between vectors is indicated with a dot, and the vector product with a cross

Therefore, the Pfaffian form

$$dx_i = g_{ij}\, dx^j = v_i \cdot dP \tag{6}$$

is in general not a perfect differential and as a consequence the parallelogrammoid defined by two displacements

$$dP = v^i dx_i\,, \qquad \delta P = v^i \delta x_i \tag{7}$$

will show a misclosure ΔP given by

$$\Delta P = (\delta d - d\delta) P = \Omega^{ij\cdot}_{\cdot\cdot h} v^h\, dx_i \delta x_j \tag{8}$$

and in general $\Omega^{ij\cdot}_{\cdot\cdot h} \neq 0$.

Contrary to the coordinates x^i, the coordinates x_i defined by Eq. (6) are therefore non-holonomic.

The non-holonomity of the coordinates x_i is highlighted if we consider the coefficients of connection between neighboring triads as defined by

$$dv_i = \begin{Bmatrix} h \\ i\ \ j \end{Bmatrix} v_h\, dx^j\,; \qquad dv^i = \begin{Bmatrix} i\ \ j \\ h \end{Bmatrix} v^h dx_j\,; \tag{9}$$

for which

$$\begin{Bmatrix} h \\ i\ \ j \end{Bmatrix} - \begin{Bmatrix} h \\ j\ \ i \end{Bmatrix} \equiv 0 \tag{10}$$

$$\begin{Bmatrix} i\ \ j \\ h \end{Bmatrix} - \begin{Bmatrix} j\ \ i \\ h \end{Bmatrix} = 2\,\Omega^{ij\cdot}_{\cdot\cdot h}\,. \tag{11}$$

As already remarked, $\Omega^{ij\cdot}_{\cdot\cdot h}$ is in general different from zero and is called the *object anholonomity* relevant to the coordinates x_i.

The symbols $\begin{Bmatrix} h \\ i\ \ j \end{Bmatrix}$ are the symbols of Christoffel of the second kind; and we obviously have

$$\begin{Bmatrix} i\ \ j \\ h \end{Bmatrix} = -\,g^{rj} \begin{Bmatrix} i \\ h\ \ r \end{Bmatrix}. \tag{12}$$

Indicating by (ij, h) the symbols of Christoffel of the first kind

$$(ij, h) = g_{hk} \begin{Bmatrix} k \\ i\ \ j \end{Bmatrix} \tag{13}$$

and putting

$$\Omega_{ij,h} = g_{ir}\, g_{js}\, \Omega^{rs\cdot}_{\cdot\cdot h} \tag{14}$$

whe also have

$$2\,\Omega_{ij,h} = (hi, j) - (hj, i)\,. \tag{15}$$

3. We shall in the following consider the differential operators curl and divergence.

As for the operator curl we have

$$\operatorname{curl} \boldsymbol{v}^{i} = \operatorname{curl} \operatorname{grad} x^{i} \equiv 0 \tag{16}$$

and from the definition of the operator

$$\operatorname{curl} \boldsymbol{v}_{h} \cdot \boldsymbol{v}_{i} \times \boldsymbol{v}_{j} = (hi, j) - (hj, i) = 2\,\Omega_{ij,h} = \frac{\partial g_{hj}}{\partial x^{i}} - \frac{\partial g_{hi}}{\partial x^{j}}\,. \tag{17}$$

We may also write

$$\operatorname{curl} \boldsymbol{v}_{h} = \frac{1}{\sqrt{G}} \begin{vmatrix} \boldsymbol{v}_{1} & \boldsymbol{v}_{2} & \boldsymbol{v}_{3} \\ \dfrac{\partial}{\partial x^{1}} & \dfrac{\partial}{\partial x^{2}} & \dfrac{\partial}{\partial x^{3}} \\ g_{h1} & g_{h2} & g_{h3} \end{vmatrix} \tag{18}$$

in which

$$G = |g_{rs}|. \tag{19}$$

We therefore have for any vector $\boldsymbol{u} = u_{i}\boldsymbol{v}^{i}$

$$\operatorname{curl} \boldsymbol{u} = \frac{\partial u_{i}}{\partial x^{j}}\, \boldsymbol{v}^{i} \times \boldsymbol{v}^{j}\,. \tag{20}$$

4. Considering the operator divergence, we have

$$\operatorname{div} \boldsymbol{v}^{i} = \left\{ \begin{matrix} i & k \\ & h \end{matrix} \right\} = \Delta_{2} x^{i} = -g^{hk} \left\{ \begin{matrix} & i & \\ h & & k \end{matrix} \right\} = \tau^{i} \tag{21}$$

and therefore

$$\operatorname{div} \boldsymbol{u} = g^{ij} \frac{\partial u_{i}}{\partial x^{j}} + u_{i}\tau^{i}\,. \tag{22}$$

Conversely,

$$\operatorname{div} \boldsymbol{v}_{i} = \left\{ \begin{matrix} & h & \\ i & & h \end{matrix} \right\} = \frac{\partial}{\partial x^{i}} \lg \sqrt{G} = S_{/i} \tag{23}$$

and therefore

$$\operatorname{div} \boldsymbol{u} = g^{ij} \frac{\partial u_{i}}{\partial x^{j}} + u^{i} S_{/i} \tag{24}$$

in which

$$S = \lg \sqrt{G}\,. \tag{25}$$

Consequently, the divergence of the base vectors, as well as the divergence of the reciprocal vectors, are the components of two vectors respectively, both characteristic of the assigned coordinate system:

– the divergence of the base vectors, the covariant components of the logarithmic gradient of the volume spanned by the base vectors themselves:

$$\operatorname{grad} S = \boldsymbol{v}^{i} \operatorname{div} \boldsymbol{v}_{i} = S_{/i} \boldsymbol{v}^{i} = \left\{ \begin{matrix} & h & \\ i & & h \end{matrix} \right\} \boldsymbol{v}^{i}\,; \tag{26}$$

– the divergence of the reciprocal vectors, the contravariant components of the displacement vector $\boldsymbol{\tau}$:

$$\boldsymbol{\tau} = \boldsymbol{v}_i \operatorname{div} \boldsymbol{v}^i = \boldsymbol{v}_i \Delta_2 x^i = \left\{ \begin{matrix} i & h \\ & h \end{matrix} \right\} \boldsymbol{v}_i \,. \tag{27}$$

It should also be noted that

$$2\Omega^{ir}_{\cdot\cdot h} = \tau^i + S^{/i}; \qquad S^{/i} = g^{ir} S_{/r} \,. \tag{28}$$

5. In the following an application is made to the case of the intrinsic coordinates $x^1 = \varphi$ (geodetic latitude), $x^2 = \lambda$ (geodetic longitude), $x^3 = U$ (normal potential) of the normal gravity field based on the standard biaxial ellipsoid for which the metric tensor is defined by

$$ds^2 = \rho^2 d^2\varphi + \nu^2 \cos^2\varphi d\lambda^2 + \frac{1+f^2}{\gamma^2} dU^2 + \frac{2\rho f}{\gamma} d\varphi dU \,. \tag{29}$$

Here ρ and ν are the radii of curvature along the meridian and the prime vertical respectively, γ the normal gravity, and

$$f = \frac{\partial \lg \gamma}{\partial \varphi} \,. \tag{30}$$

It is hardly necessary to note that the same formulae are rigorously valid for any field endowed with rotational symmetry.

Indicating by $(\boldsymbol{i}_1, \boldsymbol{i}_2, \boldsymbol{i}_3)$ the local orthonormal triad in which $\boldsymbol{i}_1$ is directed to the North, $\boldsymbol{i}_2$ to the East, and $\boldsymbol{i}_3$ to the Zenith, we have

$$\operatorname{curl} \boldsymbol{i}_1 = \frac{\boldsymbol{i}_2}{\rho}; \qquad \operatorname{curl} \boldsymbol{i}_2 = \frac{\boldsymbol{i}_1}{\nu} - \frac{\boldsymbol{i}_3 \operatorname{tg}\varphi}{\nu}; \qquad \operatorname{curl} \boldsymbol{i}_3 = f \boldsymbol{i}_2 \tag{31}$$

$$\operatorname{div} \boldsymbol{i}_1 = -\frac{\operatorname{tg}\varphi}{\nu} - \frac{f}{\rho}; \qquad \operatorname{div} \boldsymbol{i}_2 = 0; \qquad \operatorname{div} \boldsymbol{i}_3 = \mathscr{H} = \frac{1}{\rho} + \frac{1}{\nu} \tag{32}$$

and then, taking into account that

$$G = \frac{\rho^2 \nu^2 \cos^2\varphi}{\gamma^2} \tag{33}$$

and the formulae given in (Marussi, 1950) for the Christoffel's symbols and for derivatives of the coefficients of the metric tensor:

$$\begin{aligned} \operatorname{curl} \boldsymbol{v}_1 &= [2 + f(f+g) - f'] \boldsymbol{i}_2 \\ \operatorname{curl} \boldsymbol{v}_2 &= -2\cos\varphi \boldsymbol{i}_1 - 2\sin\varphi \boldsymbol{i}_2 \\ \operatorname{curl} \boldsymbol{v}_3 &= \frac{1}{\rho\gamma} \left[(1+f^2)(f+g) + ff' + \frac{\rho}{\nu}(2f+h) + \frac{6\rho f \omega^2}{\gamma} \right] \boldsymbol{i}_2 \\ \operatorname{div} \boldsymbol{v}^1 &= \Delta_2 \varphi = \tau^1 = \frac{1}{\rho} \left(\frac{2f + h - \operatorname{tg}\varphi}{\nu} + \frac{4 f \omega^2}{\gamma} \right) \end{aligned} \tag{34}$$

$$\operatorname{div} \boldsymbol{v}^2 = \Delta_2 \lambda = \tau^2 = 0 \tag{35}$$
$$\operatorname{div} \boldsymbol{v}^3 = \Delta_2 U = \tau^3 = 2\omega^2$$

$$\operatorname{div} \boldsymbol{v}_1 = S_{/1} = -\operatorname{tg}\varphi + g + h - f$$
$$\operatorname{div} \boldsymbol{v}_2 = S_{/2} = 0 \tag{36}$$
$$\operatorname{div} \boldsymbol{v}_3 = S_{/3} = \frac{-1}{\gamma}\left(2\mathscr{H} + \frac{2\omega^2}{\gamma} + \frac{2f^2 - f'}{\rho} - \frac{f \operatorname{tg}\varphi}{\nu}\right)$$

in which

$$f = \frac{\partial \lg \gamma}{\partial \varphi}, \quad g = \frac{\partial \lg \rho}{\partial \varphi}, \quad h = \left(1 - \frac{\rho}{\nu}\right) \operatorname{tg}\varphi, \quad f' = \frac{\partial f}{\partial \varphi} \tag{37}$$

are all small quantities, and ω is the angular speed of rotation of the Earth.

Bibliography

Marussi A (1950) Principi di Geodesia intrinseca applicati al campo di Somigliana. Boll Geodes Sci Affini 8 (2) (Chap. III.1, this Vol.)

Marussi A (1951) Fondamenti di Geodesia intrinseca. Comm Geodes Ital Mem no 7 (Chap. I.2, this Vol.)

Marussi A (1979) Natural reference systems and their reciprocals in geodesy. Publ Finn Geod Inst no 89

IV Mapping of the Actual Gravity Field onto the Normal Reference Field

1 On the Representation of the Geoid on the Ellipsoid [1]

1. Recently the discussion has been reopened on the problem of the representation of a geodetic network on the ellipsoid (Vening-Meinesz et al. 1950), a question whose importance is not only theoretical, but also practical, since very extensive geodetic networks of continental scale have been unified into great complexes, and still more will be in the near future. With the growth in the dimensions of those complexes, the divergence between the actual surface of the geoid and that of the ellipsoid becomes increasingly in evidence, and account must be taken of those discrepancies which could for the most part be ignored in practice when working in more limited regions.

As has already become apparent also in the recent discussions, such a question can in turn be split into two distinct problems, at least so long as one wishes to avoid placing the question squarely in its natural setting of a three-dimensional space (Vening-Meinesz et al. 1951).

The two problems are as follows:

1st problem: to establish a one-to-one correspondence between the points on the Earth's physical surface which have been the concern of the geodetic measurements, and the points of the geoid;

2nd problem: to establish a one-to-one correspondence between the points of the geoid, and the points of the reference ellipsoid. In the establishment of both these correspondences, it is as well to have in mind the ultimate objective of the procedure, which is to be able to exploit the metric of a well known surface for the rapid calculation of the geometric relationships of position between the points of the Earth.

2. It is convenient to straight away underline the profound difference which exists, from a strictly logical point of view, between the two problems. Since the correspondence between the points of the Earth's physical surface and the geoid is made in the most natural way by means of the transport of the point along the line of force passing through it, which has therefore a physical reality, then the solution of the first problem depends uniquely on the knowledge of the geometric structure of the gravity field, attainable by measurements which in Geodesy can be carried out, at least potentially. It does not, therefore, entail the introduction of supplementary and arbitrary hypotheses.

The resolution of the second problem, on the contrary, is removed from physical reality, and implies the introduction of an imaginary and arbitrary

[1] Originally published as: Marussi A (1951) Sulla rappresentazione del geoide sull'ellissoide. Boll Geodes Sci Affini 10: no 3

surface, and of an arbitrary law of correspondence. This surface is restricted solely by the practical convenience of keeping its shape as close as possible to that of the geoidal surface, and of making the correspondence, as far as possible, isometric.

3. Leaving aside the first problem, to which other studies will be devoted, we shall suppose of course that such a surface on which to represent the geoid G will be taken as an appropriate ellipsoid E. The law of correspondence between these two surfaces is conceived for the most part to be obtained "by projection" along normals (it is of minor interest to specify whether one deals with normals of the one or of the other surface) after having oriented the ellipsoid E with its minor axis along the Earth's axis, and having aligned it tangentially to one point of G (origin point or datum of geographic coordinates). It is evident from this that there is a close resemblance between this procedure and the genesis of geographic maps obtained by projection.

But already here it is appropriate to note that, from an analytical point of view, the laws of correspondence between the two surfaces, and the study of all their metric properties, depend solely on the knowledge of the first fundamental quadratic form of the surfaces themselves; while the law of correspondence "by projection" demands likewise the knowledge of the two surfaces in space, and therefore, also of their second fundamental quadratic form. This observation highlights the circumstance that the apparent simplicity, by which the correspondence between G and E is generated in this way, is not accompanied by analytic expansions of matching simplicity; and this because the elements required for their construction are excessive with respect to those which are effectively involved in the final result.

4. For this reason it appears appropriate to examine the question from a different and simpler point of view, by recourse to the general theory of the representation of surfaces on surfaces, and to proceed by exploiting only the knowledge of the surface metrics of G and E.

From the point of view of practical applications, this is by no means unimportant when one of the two surfaces, that of G in our case, has to be obtained by a process of numerical integration. While it is in fact relatively easy to know the surface metric and the total curvature of G at every point (and already the angular excesses of geodetic triangles supply, in theory, the total curvature without need of recourse to astronomical observations), the knowledge of its form in space is linked to satisfying the Mainardi-Codazzi equations, which connect the first to the second fundamental form.

5. There are already in Geodesy many examples of such analytic representations between surfaces, for various purposes, and also in the problem which has given rise to the present discussions. One such is the representation of the ellipsoid on itself, when the same triangulation network is depicted on the same ellipsoid, but with different initial orientation parameters, as Prof. Vening-Meinesz has explained in his last letter addressed to Prof. Baeschlin. The two alternatives put forward by the illustrious geodesist are equivalent, the first (Helmert, Baeschlin, Hotine) to a representation of the analytic type which can be identified with the azimuthal equidistant representation having centres in the

two source points, and the second (Vening-Meinesz) to a representation "by projection".

The question, brought up by General Hotine, of whether a change in the initial longitude should then entail a change in azimuth according to Laplace's equation, departs from the problem of the representation and is concerned with the physical aspect of the question; it always seems that the reply should be in the affirmative in every case in which a geophysical character can be attributed to such a change in longitude, which is confirmed in the great majority of cases.

Another example of representation between surfaces which might have had geodetic use is the conformal representation of the ellipsoid on the sphere, due to Gauss, and which obviously is numbered amongst the analytical representations. Among these also must be reckoned the conformal representations of the ellipsoid on the plane, which have assumed such importance in modern Geodesy.

6. These considerations lead one to ask whether it might not also be opportune to have recourse, for the representation of G on E, to correspondences of the analytical type rather than, as has been done until now, to those "by projection"; and whether it might not be opportune to apply one's attention directly to conformal projections, whose advantages are evident in this sort of question. One should then be able to note that the ordinary usage followed in the computation of a geodetic network, which consists in the most part of transferring unchanged the geodetic angles onto E, constitutes in itself the kernel of a conformal representation, insofar as the network triangles are considered to be small.

This means of presentation is modified, to tell the truth, by the subsequent adjustment of the network as usually made, by which one is obliged to seek a basis on E rather than on G, with the introduction of the ellipsoidal excesses of triangles in place of the actual excesses, and linking the network through the equations to the baselines as if these had been measured on E.

In this way one exerts a strain on the angles of the network, forcing the correspondence between E and G to be more strictly isometric, rather than purely conformal.

If it is desired on the other hand to preserve the angles, and to study the linear changes brought about by the conformal representation between G and E, it should be remembered that such changes are governed by the equation of Laborde, which can be written in quite a simple form, as has been shown elsewhere (Marussi 1951; Chap. V.1, this Vol.):

$$(\Delta_2\mu)_G = K_G - K_E\, e^{2\mu} \tag{1}$$

or

$$(\Delta_2\mu)_E = K_G\, e^{-2\mu} - K_E \tag{2}$$

in which $(\Delta_2\mu)_E$ *and* $(\Delta_2\mu)_G$ represent the second differential parameter, computed respectively on E and on G, of u, the logarithm of the modulus of deformation:

$$\mu = \lg \frac{ds_E}{ds_G} \tag{3}$$

and K_E and K_G are the total curvatures of the two surfaces at points which are in correspondence in the chosen representation.

7. Supposing that the metrics of the two surfaces are known, and hence their total curvatures, the integration of the above equation permits the derivation at every point of the modulus of linear deformation; or alternatively one can choose at will a line on one of the two surfaces (the curve bearing the boundary values) and on it the modulus itself and its normal derivative. For convenience, it is always easier to choose such a curve on E, and to decide that along it, both μ and its normal derivative shall be zero. According to whether this line is taken to coincide with the equator, with a meridian, or with a parallel, etc., one has obvious extensions of the projections of Mercator, of Gauss, of Lambert, etc.; and the defining line conserves in the representation its tangential curvature (if it is a geodesic, it therefore is transformed into a geodesic).

The serious difficulty, which is met with in the integration of Laborde's equation in the general case, is that the total curvatures K_E and K_G should be referred to corresponding points in the representation, but this itself is unknown. In other words, with the introduction of coordinates, for example on E, then K_E will be known as a function of these coordinates, but K_G will not, since the links of correspondence between E and G have not yet been established. But in our real case, because of the rather slow way in which the difference between K_E and K_G varies, whatever may be the quasi-isometric correspondence finally established between E and G, both total curvatures can be regarded as known as functions of the coordinates adopted on one of the two surfaces, in our case on E.

The integration of Laborde's equation is then easily carried out in a single operation, with the chosen boundary conditions.

8. In the sequence of ideas briefly illustrated here, the procedure for computation of triangulations on E should be as follows. It is necessary first of all to determine at every point of the geoid its total curvature, in such a way as to allow it to be shown graphically, for example, on a map of lines of equal value. As already noted, knowledge of the total curvature can be had quite rigorously, at least within the limits of observational errors, solely from surface measurements of the spheroidal excesses; it can, however, be determined with high precision by means of astronomic-geodetic measurements, and with requirements much less taxing than those needed by the method "by projection", which demands the complete knowledge of the surface in space.

Having done this, and computed to a first approximation the triangulation network on the geoid (i.e., closing the triangles by using the geoidal angular excess, and imposing an exact restitution of the geodetic baselines reduced to the geoid) one can then proceed to the integration of Laborde's equation, taking as the curve carrying the boundary values $\left(\mu = 0 \; , \dfrac{d\mu}{dn} = 0\right)$ a curve which is well adapted to the region of interest, or even more simply a meridian or a parallel. This will supply point by point the value of the modulus of deformation which can be taken into account, as is done in the ordinary use of the conformal plane in geodesy, through the transfer of the network onto the ellipsoid by way of the reduction of the baselines and the computations of the "reductions to the chord" (in our case "reductions to the ellipsoidal geodesic").

9. In what has been said up to this point, it has not yet been made clear in what manner one can take account, in the above-mentioned procedure, of the Laplace equation, or of the analogue which might in the end need to be substituted for it; the discussion of this aspect of the problem, which today one can only sketch briefly, bears on questions both general and fundamental, which re-open the old controversy on the possibility of determining longitudes by means of the knowledge of only latitudes and of the convergence of the astronomic azimuths of the reciprocal normal sections – a question that was of the highest importance in the days before the discovery of radio.

Here the problem appears as naturally inverted, in so far as from the knowledge of the longitude one seeks to recover the forementioned convergence; but its substance remains unchanged.

Without wishing to enter into detail on the difficult question as to whether the application of the Laplace equation as made in ordinary use is more or less rigorous, we shall confine ourselves to the observation that it allows us to deduce – excepting of course the observational errors – the differences of astronomical longitude, which are absolute quantities independent of the reference ellipsoid, from other quantities which are derived solely by observation. The ellipsoid appears in this process as a simple intermediary, without leaving the slightest trace on the result.

This simple observation betrays the intrinsic significance of Laplace's equation, which should therefore be capable of enunciation in terms independent of either the ellipsoid or of the manner in which the geoid is represented on it.

Now this enunciation is perhaps contained in the famous theorem of Dalby, which relates the convergence of the normal sections at two points of an ellipsoid with the convergence of the meridians of the corresponding image on the sphere of the directions, and affirms that the two convergences differ by quantities of the 5th order of smallness. The generalization of Dalby's theorem for any surface whatever relates this difference, as I undertake to show elsewhere, to the concomitant effect of the geodetic torsion and of the variability of the radius of curvature in the direction which joins the two points considered; and because of this the difference remains rather small even when one wishes to consider this fundamental theorem as applying to the geoid rather than to the ellipsoid.

It was in this way that the problem came to be put in the past, even though its application was directed to obtaining the longitudes from the azimuths; and I should like to recall here the words of Ivory (1828) in his controversy with Tiarks (1828), which confirm how the value of this theorem may have been seen, even in the mind of that illustrious geodesist, as standing outside the laws of the conventional representation of the geoid by means of ellipsoidal geodesics:

"...We have advanced no vague reasoning about the well-known properties of spheroidical triangles and geodetical lines, in a case where, in fact, there is neither any such triangle nor any such line."

And further

"...But he has entirely mistaken the nature of the problem and the difficulties that must be overcome in solving it. There is no spheroidical triangle; there is no geodetical line. The data consist of the latitudes, nearly equal, and the azimuths, of two distant stations, yet sufficiently near for one to be seen from the other." (Ivory 1828, pp 434 – 5).

Bibliography

Ivory J (1828) On the method in the trigonometrical survey for finding the difference of longitude of two stations very little different in latitude. Philos Mag, Ann Philos, December 1828:432

Marussi A (1951) Su alcune proprietà fondamentali delle rappresentazioni conformi fra superfici qualunque. Boll Geodes Sci Affini no 2

Tiarks Dr. (1828) On Mr Dalby's method of finding the difference of longitude between two points of a geodetical line on a spheroid, from the latitude of those points and the azimuths of the geodetical line at the same. Philos Mag, Ann Philos, November 1828:364

Vening-Meinesz FA, Baeschlin F, Hotine M (1950) New formulas for systems of deflections of the plumb-line and Laplace's theorem. Bull Géodés no 11

Vening-Meinesz FA, Baeschlin F, Hotine M (1951) Deviations of the vertical. Discussion. Emp Surv Rev April 1951:69

2 Some Remarks on the Use of Conformal Representations in Three-Dimensional Geodesy [1]

1. A tribute as such should not be a thing of practical use, but rather should exhibit an aesthetic attraction. And since I wish to offer such a tribute to my colleague Grossmann on the occasion of his 70th birthday, I have thought that a theme from *Geodesia Intrinseca* (Intrinsic Geodesy) might fulfil this aim. I have devoted myself for so many years to this subject, but its latest developments have eluded me and have been followed up by other distinguished colleagues.

In my opinion there can be no doubt whatever that this subject exerts an aesthetic attraction: in *Geodesia Intrinseca* we seek to rid our discipline of all the prejudices and artificial constructions that have accumulated in the course of the centuries, and also to loosen it from all the bonds of practical necessity to which it has all too often become subjected. It is sought to fashion this study only on the basis of such quantities as possess a physical reality, and thereby to discern the important from the trivial, and unity from diversity.

It is equally my opinion that this subject is of no practical use, at least to judge by the present state of things and by considering immediately practical needs; perhaps this is not quite the case if we take account of the stimulus and knowledge which could arise from this discipline, and which in part has already arisen.

2. The world studied by Geodesy is a complicated world, which in various ways we have wished to describe in a simple form, by which it becomes accessible to our imagination. Unfortunately it has happened and still often happens today, that by the use of arbitrary simplifying, reducing, and normalizing processes, the very object of our study is distorted, and thus all deductions are prejudiced. Because of this it is particularly noteworthy that very recently there has been an effort, due principally to Martin Hotine, to fashion a simpler, synthetic picture of this geodetic world, without, for this reason, deforming it.

The underlying concept here is very simple, and the cartographer is especially committed to it, since he represents the figure appearing on one surface on another surface differing from the first, in order to bring into prominence the particular properties of these figures.

3. Let us now consider in a three-dimensional space, the family of equipotential surfaces $W = \text{const}$ of the gravity field, in which we recognize that neither are they mutually equidistant, nor are the lines of force geodesic lines; and the question

[1] Originally published as: Marussi A (1967) Einige Bemerkungen über die Anwendung der konformen Abbildungen in der dreidimensionalen Geodäsie. In: Aus der geodätischen Lehre und Forschung – Festschrift zum 70. Geburtstag von Walter Grossmann. Witwer, Stuttgart

now immediately arises, whether it might not be possible to suitably represent the gravity field in another space, so that such desired properties of these surfaces and curves would be realized in the representation.

This can be attained by a conformal representation of the Euclidean space V_3, in which our gravity field lies, onto another three-dimensional space $\bar{V}_3$. This space proves to be curved in Riemann's sense. The scale factor that governs the representation is then identical with the value of gravity g (Hotine 1965a).

I may now be permitted to make a few remarks on this theme, particularly on the variations which the curvatures undergo in such a representation.

4. Let us refer V_3 and $\bar{V}_3$ to the same coordinate system $\varkappa^i$ $(i=1,2,3)$ and thereby ensure a reciprocal and unequivocal correspondence, and let a_{rs} and $\bar{a}_{rs}$ be the metric fundamental tensors of V_3 and $\bar{V}_3$ respectively, so that it follows from the conformality of the representation that

$$\bar{a}_{rs} = e^{2\gamma} a_{rs}, \qquad \gamma = \ln g, \qquad g = |\operatorname{grad} W|. \tag{1}$$

Now let Σ be a surface of V_3, whose equation is $\varkappa^i = \varkappa^i(u^1, u^2)$, and $\bar{\Sigma}$ its transform in $\bar{V}_3$. If the second fundamental tensor of Σ is denoted by $b_{\alpha\beta}(\alpha, \beta, \ldots = 1, 2)$[1] $\left(\text{of which the first is } a_{\alpha\beta} = a_{ij}\varkappa^i_\alpha \varkappa^j_\beta,\ \varkappa^i_\alpha = \dfrac{\partial \varkappa^i}{\partial u^\alpha}\right)$, then its transform becomes (cf. Bianchi 1930, p. 467)

$$\bar{b}_{\alpha\beta} = e^{\gamma}(b_{\alpha\beta} - a_{\alpha\beta}\gamma_n) \tag{2}$$

where $\gamma_n = \dfrac{\partial \gamma}{\partial n}$ is the derivative of γ in the direction of the normal $\boldsymbol{n}$ to Σ.

If we observe a surface direction λ^α on Σ (and it is immediately established that $\bar{\lambda}^\alpha = e^{-\gamma}\lambda^\alpha$), then the curvature $1/r$ of the normal section for Σ in this direction is determined as $1/r = b_{\alpha\beta}\lambda^\alpha\lambda^\beta$, and correspondingly for $\bar{\Sigma}$; from this we have straight away

$$1/\bar{r} = e^{-\gamma}(1/r - \gamma_n)$$

where the change of curvature is given by the term $e^{-\gamma}\gamma_n$, which is equal to the specific flux of the logarithm of the scale factor across the curve on its plane. This is the well-known formula of Schols which, as has been demonstrated, is valid also for three-dimensional space (Hotine 1965b).

The mean curvature H of Σ is determined by the linear invariant of $b_{\alpha\beta}$

$$H = 1/r_1 + 1/r_2 = a^{\alpha\beta} b_{\alpha\beta}; \tag{3}$$

from which it immediately follows that

$$\bar{H} = e^{-\gamma}(H - 2\gamma_n); \tag{4}$$

and its relative total curvature is found from the quadratic invariant

$$K = 1/r_1 r_2 = \tfrac{1}{2}\varepsilon^{\alpha\beta}\varepsilon^{\delta\eta} b_{\alpha\delta} b_{\beta\eta} \tag{5}$$

where $\varepsilon^{\alpha\beta}$ is Ricci's tensor; hence it follows that

$$\bar{K} = e^{-2\gamma}(K - H\gamma_n + \gamma_n^2). \tag{6}$$

[1] In the following, all Roman indices run from 1 to 3, and Greek from 1 to 2

5. It is clear that the second fundamental tensor $\bar{b}_{\alpha\beta}$, together with the metric tensor, describes the form of the surface in relation to the surrounding space. On this basis the expression $1/r_1 r_2$ has been called the relative total curvature. We shall come to another result, if we by contrast estimate the total curvature of the surface, which we shall call in this case the Gaussian or Riemannian curvature, with the help of the Riemann-Christoffel tensor appropriate to it. As is well known, this is expressed through the fundamental tensor of the surface only.

We thus have (Bianchi 1930, p. 452; Weatherburn 1950, p. 138)

$$\bar{b}_{\alpha\beta}\bar{b}_{\mu\nu}-\bar{b}_{\alpha\mu}\bar{b}_{\beta\nu}=\bar{R}_{\alpha\mu\beta\nu}-\bar{R}_{ijhk}\varkappa^i_\alpha\varkappa^j_\beta\varkappa^h_\mu\varkappa^k_\nu \tag{7}$$

so that, if $\alpha=\beta=1$, $\mu=\nu=2$,

$$\bar{K}=1/\bar{r}_1\bar{r}_2=\bar{\mathfrak{K}}-\bar{K}_0 \tag{8}$$

where $\bar{K}_0$ is the Riemannian curvature of the surrounding space $\bar{V}_3$, in the direction of the surface considered.

The foregoing equation can be written as follows

$$\bar{\mathfrak{K}}=\bar{K}+\bar{K}_0 \tag{9}$$

and $\bar{\mathfrak{K}}$ is thus called the absolute total curvature of the surface.

6. We now establish an extremely important fact. The change in curvature $e^\gamma\gamma_n$ depends only on the position and on the direction of the normal to the surface, and not on the direction of the section on it. Since it is the same for all these directions, it follows therefore that the directions of principal curvature are conserved in the representation. Thus if a family of surfaces is not of Lamé's type in V_3, i.e., if it does not belong to a triply-orthogonal system, then neither can its transform in $\bar{V}_3$. Since the family of equipotential surfaces of the Earth's gravity field is not of Lamé's type (for it to be so, Darboux' third order partial differential equations would need to be satisfied), then neither can be any of its transforms in a conformal representation; there is thus no possibility of reducing the study of the Earth's potential field onto a triply-orthogonal coordinate system.

7. It has already been remarked that the conformal representation of the Euclidean space V_3 onto the Riemannian space $\bar{V}_3$ leads to the curvatures of the latter, which are described in a general form by the Riemann-Christoffel tensor $\bar{R}^{j\cdot}_{i\,kr}$. This is related to the tensor $R^{j\cdot}_{i\,kr}$ (identically zero) of V_3 by the equation

$$\bar{R}^{j\cdot}_{i\,kr}=R^{j\cdot}_{i\,kr}+\delta^j_k\Gamma_{ir}-\delta^j_r\Gamma_{ik}+a^{jh}(a_{ir}\Gamma_{kh}-a_{ik}\Gamma_{rh})-(\delta^j_r a_{ki}-\delta^j_k a_{ri})\Delta_1\gamma \tag{10}$$

where $\Gamma_{ir}=\gamma_{ir}-\gamma_i\gamma_r$, and γ_i and γ_{ir} are the tensor derivatives of γ, δ^j_k is the Kronecker symbol, and $\Delta_1\gamma=(\operatorname{grad}\gamma)^2$.

It is also well known that all the properties of the Riemann-Christoffel tensor are contained in a contracted tensor which is associated with the former, and is also known as Ricci's tensor $R_{ir}=R^{k\cdot}_{i\,kr}$; and we have (see also Hotine 1965a)

$$\bar{R}_{ir}=R_{ir}+\Gamma_{ir}+a_{ir}(\Delta_1\gamma+\Delta_2\gamma)\,, \tag{11}$$

where Δ_2 is the Laplacian operator (Eisenhart 1949, p. 90).

With a further contraction we have the curvature scalar

$$\bar{R} = e^{-2\gamma}(R + 4\Delta_2\gamma + 2\Delta_1\gamma) . \tag{12}$$

Along with the contracted Riemannian tensor we shall consider the *curvature tensor* given by

$$S_{ir} = R_{ir} - \tfrac{1}{2} R a_{ir} , \tag{13}$$

whose linear invariant $S = a^{ir} S_{ir}$, is known as the mean curvature of the space; and taking account that $R = -2S$ we have

$$\bar{S} = e^{-2\gamma}(S - 2\Delta_2\gamma - \Delta_1\gamma) , \tag{14}$$

and hence $\bar{S}_{ir} = S_{ir} + \Gamma_{ir} - a_{ir}\Delta_1\gamma$.

8. The curvature tensor allows the Riemannian curvature K_0 of the space to be computed for any chosen plane by the formula $\bar{K}_0 = \bar{S}_{ir}\bar{v}^i\bar{v}^r$ where v^i is the contravariant component of the normal to the plane, whence

$$\bar{K}_0 = e^{-2\gamma}(\gamma_{nn} - \gamma_n^2 - \Delta_2\gamma) \tag{15}$$

since obviously $K_0 = 0$, and provided that $\gamma_{nn} = \gamma_{ij} v^i v^j$.

Now $\bar{\mathfrak{K}}$, the absolute curvature of $\bar{\Sigma}$, is the sum of its relative curvature $\bar{K}$ and of the Riemannian curvature $\bar{K}_0$, so that

$$\bar{\mathfrak{K}} = \bar{K} + \bar{K}_0 = e^{-2\gamma}(K - \Delta_2\gamma + \gamma_{nn} - H\gamma_n) . \tag{16}$$

Moreover, since $\Delta_2\gamma + H\gamma_n - \gamma_{nn} = (\Delta_2\gamma)_\Sigma$, where $(\Delta_2)_\Sigma$ is the Laplacian operator on Σ, we have

$$\bar{\mathfrak{K}} = e^{-2\gamma}[K - \Delta_2\gamma)_\Sigma] .$$

Bibliography

Bianchi L (1930) Lezioni di geometria differenziale, vol II, part 2, 3rd edn. Zanichelli, Bologna

Eisenhart LP (1949) Riemannian geometry. Princeton Univ Press, Princeton

Hotine M (1965a) Geodetic applications of conformal transformations in three dimensions. Symp Math Geodes, Torino

Hotine M (1965b) Triply orthogonal coordinate systems. Symp Math Geodes, Torino

Marussi A (1951a) Su alcune proprietà fondamentali delle rappresentazioni conformi fra superfici qualunque. Boll Geodes Sci Affini X:2

Marussi A (1951b) Sulla rappresentazione del geoide sull'ellissoide. Boll Geodes Sci Affini X:3 (Chap. IV.1, this Vol.)

Marussi A (1951c) Su alcune proprietà integrali delle rappresentazioni conformi di superfici su superfici. Rend Accad Naz Lincei: Ser 8, X: fasc 4 (Chap. V.1, this Vol.)

Marussi A (1953a) Un'analogia fra le leggi della propagazione della luce in mezzi rifrangenti continui isotropi, e le rappresentazioni conformi. Atti Fond Giorgio Ronchi, anno VIII: no 1 (Chap. VI.1, this Vol.)

Marussi A (1953b) Sulla riduzione regionale e sul metodo delle derivate seconde in gravimetria. Ann Geofis 6: no 2

Marussi A (1957) Sulle rappresentazioni fra superfici definite mediante la forma quadratica che ne determina il modulo di deformazione. (Communications 75th birthday) CF Baeschlin, Zürich (Chap. V.3, this Vol.)

Weatherburn CE (1950) An introduction to Riemannian geometry and the tensor calculus. Cambridge Univ Press, Cambridge

3 On the Three-Dimensional Computation of Geodetic Networks and Anomalies[1]

List of Symbols

W	actual potential of Earth;
Φ, Λ	astronomical latitude and longitude;
$\mathbf{g}, g$	gravity vector and gravity;
U	normal potential;
φ, λ	normal or geodetic latitude and longitude;
$\boldsymbol{\gamma}, \gamma$	normal gravity vector and gravity;
T	anomalous potential;
$\tau = \dfrac{T}{U}$	relative anomaly in the potential;
$\Delta g = g - \gamma_a$	gravity anomaly;
$\mathfrak{z} = \dfrac{\Delta g}{\gamma}$	relative anomaly in gravity;
$\xi, \eta = \varepsilon \cos\varphi$	deflections of the vertical;
R	radius of curvature of line of force;
r	mean radius of Earth;
$\underset{P}{\gamma_{ij}}$	normal metric tensor evaluated at P;
$\mathbf{i}_r$	orthonormal geodetic triad in the field U ($\mathbf{i}_1$ Northw., $\mathbf{i}_2$ Eastw., $\mathbf{i}_3$ Zenith);
A, Z	astronomical azimuth and zenith distance;
α, z	normal or geodetic azimuth and zenith distance;
ρ, N	radius of curvature of meridian and of normal section perpendicular to it;
$f = \dfrac{\partial \lg \gamma}{\partial \varphi} =$	$(\lg \gamma)_{/\varphi}$;
ω	angular speed of rotation of Earth;
$T_{/r} = \dfrac{\partial T}{\partial x^r}$;	
δ^i_j	Kronecker's delta.

1. In some problems of Physical Geodesy, e.g. in Molodensky's problem, it is usual to compare measurable geometrical or physical quantities pertaining to the

[1] Originally published as: Marussi A (1973) On the three-dimensional computation of geodetic networks and anomalies. Rend Accad Naz Lincei 55: Ser 8, fasc 5, 1 – 7

actual gravity field W of the Earth, with the homologous ones in a mathematically defined reference field U, e.g., the "normal" ellipsoidal field according to Somigliana-Pizzetti's theory as adopted by the International Association of Geodesy, or the "Standard Earth" field as obtained from satellite Geodesy. The prerequisite for such comparison is the establishment of a one-to-one correspondence between points P of the field W and points Q of the field U. It is convenient to chose the point Q at an appropriate location on the equipotential surface $U = W^0$, where W^0 is the potential of the actual level surface $W = W^0$ on which P is located. The final choice of the point Q on $U = W^0$ is afterwards somewhat arbitrary, and can be best made by minimizing geometric distorsions in the mapping procedure. If point P belongs to the actual physical surface of the Earth, then point Q belongs by definition to the "telluroid" (Hirvonen 1960, 1961).

After this preliminary mapping operation has been done, all metric properties of geometric figures can be obtained by using the metric tensor pertaining to the normal field. The procedure is therefore the generalization in three dimensions of the one currently used in two dimensions in the conventional representation of the geoid onto the ellipsoid, and might be applied in the actual computation of triangulations and trilaterations in space, with the advantage of avoiding the separate treatment of planimetric and altimetric coordinates.

In the following, the properties of the correspondence are given for infinitesimal elements attached to the corresponding points P and Q; the properties for finite elements as needed in the computation of geodetic networks can afterwards be obtained e.g. by generalizing Legendre's expansions, as shown in Marussi (1950b).

The fundamental equation of Physical Geodesy, the generalized Villarceau conditions, and the harmonicity condition for the anomalous potential are also explicitly written.

2. Assume that the fields W and U are superposed in such a way that their centres of mass and their rotational axes coincide, and call $\varphi(P)$, $\lambda(P)$ and $U(P)$ (latitude, longitude, potential) the "normal" or "geodetic" coordinates of P in the field U, whereas $\Phi(P)$ $\Lambda(P)$ and $W(P)$ the "astronomical" latitude and longitude, and the actual potential, are the intrinsic coordinates of the same point in the field W. We therefore have

$$\Phi = \varphi + \xi\,, \qquad \Lambda = \lambda + \varepsilon\,, \qquad W = U + T \tag{1}$$

in which ξ and $\varepsilon = \dfrac{\eta}{\cos\varphi}$ are by definition the deflections in latitude and longitude, and T the anomalous potential.

As illustrated in a previous paper (Marussi 1973), the correspondence between P and Q will be defined by the following equations

$$\varphi(Q) = \varphi(P)\,, \qquad \lambda(Q) = \lambda(P)\,, \qquad U(Q) = W(P) = U(P) + T(P)\,; \tag{2}$$

such correspondence is therefore deflections- and potential-invariant, and it might be interpreted geometrically as the projection of point P onto point Q along the "isozenithal" line of the field U. The projection along the line of force as usually considered leads to more complicated relations than Eq. (2), but it should be noted that the difference in position resulting from the two procedures is very slight, being of the order of $\frac{T}{\gamma R} = \tau \frac{r}{R}$.

In the following the metric properties of the representation will be derived, and the relations between the adopted coordinates and measurable quantities will be given. We assume as measurables the geopotential W and g (as obtained by leveling and gravity observations), the elementary displacement $dP = \boldsymbol{t}\,ds$, where the unit vector $\boldsymbol{t}$ is observed by means of its actual astronomical azimuth A and zenith distance Z and ds by distance measuring instruments or triangulation procedures, and finally the astronomical latitude and longitude, Φ and Λ.

3. Indicating by $x^i (x^1 = \varphi,\ x^2 = \lambda,\ x^3 = U)$ the geodetic coordinates of P in the field U, and similarly by y^i the geodetic coordinates of Q, we have for a displacement $dP = \boldsymbol{t}\,ds$ and by Eq. (2) for its image $dQ = \boldsymbol{t}^* ds^*$ the following contravariant components:

$$dP \equiv dx^i = \lambda^i ds\,; \qquad \lambda^i = \frac{dx^i}{ds}\,; \tag{3'}$$

$$dQ \equiv dy^i = (\delta^i_r + \delta^i_3 T_{/r})\,dx^r = \lambda^{i*} ds^*\,; \tag{3''}$$

$$\lambda^{i*} = \frac{dy^i}{ds^*} = \frac{ds}{ds^*}(\delta^i_r + \delta^i_3 T_{/r})\,\lambda^r \tag{3'''}$$

since $dW = dU + dT = dU + T_{/r}\,dx^r$, and the summation convention is used throughout.

Indicating by $\underset{P}{\gamma_{ij}}$ and $\underset{Q}{\gamma_{ij}}$ the components of the metric tensor of the "normal" field evaluated at P and at Q respectively, we than have

$$ds^2 = \overline{dP^2} = \underset{P}{\gamma_{ij}}\, dx^i dx^j\,; \qquad \underset{P}{\gamma_{ij}} \lambda^i \lambda^j = 1\,; \tag{4'}$$

$$ds^{*2} = \overline{dQ^2} = \underset{Q}{\gamma_{ij}}\, dy^i dy^j\,; \qquad \underset{Q}{\gamma_{ij}} \lambda^{i*} \lambda^{j*} = 1\,; \tag{4''}$$

$$m^2_\lambda = \left(\frac{ds^*}{ds}\right)^2 = a_{rs} \lambda^r \lambda^s \tag{4'''}$$

in which

$$a_{rs} = \underset{Q}{\gamma_{ij}} (\delta^i_r + \delta^i_3 T_{/r})(\delta^j_3 T_{/s})\,. \tag{5}$$

Let us now write

$$\boldsymbol{t} = \frac{dP}{ds} = \boldsymbol{i}_1 \cos\alpha \sin z + \boldsymbol{i}_2 \sin\alpha \sin z + \boldsymbol{i}_3 \cos z \tag{6}$$

where $\boldsymbol{i}_r$ is the orthonormal geodetic triad at P in the field U. By the definition of the correspondence between P and Q, such a triad is identical with that at Q.

Here α and z are the "normal" or "geodetic" azimuth and zenith distance of the displacement dP; if

$$A = \alpha + \Delta\alpha$$
$$Z = z + \Delta z \tag{7}$$

are the measurable (astronomical) azimuth and zenith distance referred to the actual vertical, we have, by straightforward algebra

$$\begin{aligned} \Delta\alpha &= \varepsilon \sin\varphi + (\xi \sin A - \varepsilon \cos\varphi \cos A) \operatorname{ctg} Z\,, \\ \Delta z &= -\xi \cos A - \varepsilon \cos\varphi \sin A \end{aligned} \tag{8}$$

and therefore

$$\begin{aligned} \boldsymbol{t} = \boldsymbol{i}_1 &(\cos A \sin Z + \xi \cos Z + \varepsilon \sin\varphi \sin A \sin Z) \\ &+ \boldsymbol{i}_2 [\sin A \sin Z + \varepsilon (\cos\varphi \cos Z - \sin\varphi \cos A \sin Z)] \\ &+ \boldsymbol{i}_3 (\cos Z - \xi \cos A \sin Z - \varepsilon \cos\varphi \sin A \sin Z)\,. \end{aligned} \tag{6'}$$

4. From this point on, we shall assume the normal ellipsoidal field as the U reference field; we have (Marussi 1950d)

$$\gamma_{ij} \equiv \left\| \begin{matrix} \rho^2 & 0 & \dfrac{\rho f}{\gamma} \\ 0 & N^2 \cos^2\varphi & 0 \\ \dfrac{\rho f}{\gamma} & 0 & \dfrac{1+f^2}{\gamma^2} \end{matrix} \right\| \tag{9}$$

and for the non-zero Christoffel's symbol of the second kind:

$$\begin{aligned} &\begin{Bmatrix} 1 \\ 11 \end{Bmatrix} = \left(\lg \frac{\rho}{\gamma}\right)_{/\varphi}; \quad \begin{Bmatrix} 1 \\ 22 \end{Bmatrix} = \frac{N \cos^2\varphi}{\rho} (\operatorname{tg}\varphi - f)\,; \quad \begin{Bmatrix} 2 \\ 12 \end{Bmatrix} = -\frac{\rho}{N} \operatorname{tg}\varphi\,; \\ &\begin{Bmatrix} 1 \\ 13 \end{Bmatrix} = (\lg \rho)_{/U} \cong \frac{-1}{U}; \quad \begin{Bmatrix} 2 \\ 23 \end{Bmatrix} = (\lg N)_{/U} \cong \frac{-1}{U}; \\ &\begin{Bmatrix} 1 \\ 33 \end{Bmatrix} = \frac{1}{\gamma\rho} f_{/U} \cong \frac{1}{U} f_{/U}; \quad \begin{Bmatrix} 3 \\ 11 \end{Bmatrix} = \gamma\rho \cong U; \quad \begin{Bmatrix} 3 \\ 22 \end{Bmatrix} = \gamma N \cos^2\varphi \cong U \cos^2\varphi; \\ &\begin{Bmatrix} 3 \\ 33 \end{Bmatrix} = -(\lg \gamma)_{/U} \cong \frac{-2}{U}\,. \end{aligned} \tag{10}$$

In deriving the expressions of Christoffel's symbols, use has been made of the following relations:

$$\begin{aligned} &(\lg \rho)_{/U} = \frac{-1}{\gamma\rho} (1 + f^2 - f_{/\varphi}) \cong \frac{-1}{U}; \quad (\lg N)_{/U} = \frac{-1}{\gamma N} (1 + f \operatorname{tg}\varphi) \cong \frac{-1}{U}; \\ &(\lg N)_{/\varphi} = \left(1 - \frac{\rho}{N}\right) \operatorname{tg}\varphi\,; \quad (\lg \gamma)_{/U} = \frac{1}{\gamma} \left(\frac{1}{N} + \frac{1+f^2}{\rho} + \frac{2\omega^2}{\gamma}\right) \cong \frac{2}{U}\,. \end{aligned} \tag{11}$$

It should be noted that consequently, ignoring terms of higher order, the components of the metric tensor $\underset{P}{\gamma_{ij}}$ at P are expressed in terms of the components of $\underset{Q}{\gamma_{ij}}$ at Q by the following relations

$$\gamma_{11}_{P} = \gamma_{11}_{Q}(1+2\tau)\,; \qquad \gamma_{22}_{P} = \gamma_{22}_{Q}(1+2\tau)\,;$$

$$\gamma_{13}_{P} = \gamma_{13}_{Q}(1+3\tau)\,; \qquad \gamma_{33}_{P} = \gamma_{33}_{Q}(1+4\tau) \tag{12}$$

since the contravariant components of the displacement vector $Q-P$ are $\Delta x^i = \delta_3^i T$.

The contravariant components $\lambda^i = \dfrac{dx^i}{ds}$ of t at P are then given by the following formulae:

$$\lambda^1 = \frac{d\varphi}{ds} = \frac{l^1}{\rho_P},$$

$$\lambda^2 = \frac{d\lambda}{ds} = \frac{l^2}{N_P \cos\varphi},$$

$$\lambda^3 = \frac{dU}{ds} = -\gamma_P l^3 \tag{13}$$

in which the non-dimensional coefficients l^i are given by

$$l^1 = \cos A \sin Z + f \cos Z + \xi(\cos Z - f \cos A \sin Z) + \varepsilon \sin A \sin Z(\sin\varphi - f\cos\varphi)$$

$$l^2 = \sin A \sin Z + \varepsilon(\cos\varphi \cos Z - \sin\varphi \cos A \sin Z)$$

$$l^3 = \cos Z - \xi \cos A \sin Z - \varepsilon \cos\varphi \sin A \sin Z\,. \tag{13'}$$

Substituting in Eq. (4‴) we then have

$$a_{rs} \equiv \left\| \begin{matrix} \rho^2 & 0 & \dfrac{\rho(f-\xi)}{\gamma} \\[2ex] 0 & N^2\cos^2\varphi & \dfrac{-N\eta\cos\varphi}{\gamma} \\[2ex] \dfrac{\rho(f-\xi)}{\gamma} & \dfrac{-N\eta\cos\varphi}{\gamma} & \dfrac{1+2\mathfrak{z}+4\tau}{\gamma^2} \end{matrix} \right\| \tag{14}$$

since (Marussi 1973)

$$T_{/\varphi} = -\gamma\rho\xi \cong -U\xi\,, \qquad T_{/\lambda} = -\gamma N\eta\cos\varphi \cong -U\eta\cos\varphi\,.$$

$$T_{/U} = \mathfrak{z} + 2\tau \tag{15}$$

and

$$m^2 = \left(\frac{ds^*}{ds}\right)^2 = 1 - 2\tau\sin^2 Z$$

$$+ 2\mathfrak{z}\cos^2 Z + 2[(\xi - f)\cos A + \eta\sin A]\sin Z\cos Z \tag{15'}$$

$$m = \frac{ds^*}{ds} = 1 - \tau\sin^2 Z + \mathfrak{z}\cos^2 Z + [(\xi - f)\cos A + \eta\sin A]\sin Z\cos Z\,. \tag{15''}$$

It might be noted that in the correspondence horizontal displacements are contracted in the ratio $(1-\tau)$, and isozenithal displacements are expanded in the ratio $(1+\mathfrak{z})$. The horizontal contraction corresponds to the projection of the geoid onto the ellipsoid in the classical approach, and the vertical dilatation to the adoption of normal heights.

5. Let us now consider the vector of gravity $\boldsymbol{g} = -g\boldsymbol{i}_3$; we have by definition

$$\boldsymbol{g} = -\underset{Q}{\gamma}[\xi \boldsymbol{i}_1 + \eta \boldsymbol{i}_2 + (1+\mathfrak{z})\boldsymbol{i}_3] \,; \tag{16'}$$

$$\Delta \boldsymbol{g} = \boldsymbol{g} - \underset{Q}{\boldsymbol{\gamma}} = -\underset{Q}{\gamma}(\xi \boldsymbol{i}_1 + \eta \boldsymbol{i}_2 + \mathfrak{z}\boldsymbol{i}_3) \,; \tag{16''}$$

$$\Delta g = g - \underset{Q}{\gamma} = \underset{Q}{\gamma}\mathfrak{z} \,; \qquad g = (1+\mathfrak{z})\,\gamma \tag{16'''}$$

since $\underset{Q}{\boldsymbol{\gamma}} = -\underset{Q}{\gamma}\boldsymbol{i}_3$ because of the deflection-invariance of the correspondence between P and Q.

It has been shown in a previous paper (Marussi 1973) that the following relations hold, here written in their most concise form:

$$\mathfrak{z} = \frac{\partial T}{\partial U} - T\frac{\partial \lg \gamma}{\partial U} + T\left(\frac{1}{\gamma \rho}\,\frac{\partial f}{\partial U}\,\frac{\partial T}{\partial \varphi} - \frac{\partial \lg \gamma}{\partial U}\,\frac{\partial T}{\partial U}\right) + f\xi \tag{17}$$

$$\frac{\partial \gamma \rho \xi}{\partial \lambda} = \frac{\partial \gamma N \eta \cos\varphi}{\partial \varphi}$$

$$\frac{\partial(\mathfrak{z} - f\xi)}{\cos\varphi\, \partial \lambda} = -\gamma \frac{\partial N \eta}{\partial U}$$

$$\frac{\partial(\mathfrak{z} - f\xi)}{\partial \varphi} = -\gamma \frac{\partial \rho \xi}{\partial U} - T\frac{\partial f}{\partial U}\,. \tag{18}$$

With the approximations indicated in Eq. (11), the preceding formulae might also be written as follows

$$\frac{\partial T}{\partial U} = \mathfrak{z} + 2\tau \qquad \text{or} \qquad \mathfrak{z} = \gamma \frac{\partial}{\partial U}\left(\frac{T}{\gamma}\right) \tag{17'}$$

$$\frac{\partial \xi}{\partial \lambda} = \frac{\partial \eta \cos\varphi}{\partial \varphi}$$

$$\frac{\partial \mathfrak{z}}{\partial \varphi} = -U^2 \frac{\partial}{\partial U}\left(\frac{\xi}{U}\right)$$

$$\frac{\partial \mathfrak{z}}{\cos\varphi\, \partial \lambda} = -U^2 \frac{\partial}{\partial U}\left(\frac{\eta}{U}\right). \tag{18'}$$

Formulae (17) and (17′) express the fundamental equation of physical geodesy; the first of formulae (18) and (18′) gives the condition first established by Villarceau for the existence of the equipotential surfaces, and the last two generalise such existence condition to the equi-latitudinal and the equi-longitudinal surfaces.

In a private communication, Prof. T. Krarup has recalled my attention on the fact the harmonicity of T should also be accounted for; Laplace's equation

$$\Delta_2 T = \frac{1}{\sqrt{\Gamma}} \frac{\partial}{\partial x^i} (\sqrt{\Gamma} \gamma^{ik} T_{/k}) = 0 \,, \qquad \Gamma = | \gamma_{ij} | \tag{19}$$

is easily written in explicit form, and leads with the usual approximations to the following condition

$$\frac{\partial \xi \cos\varphi}{\partial \varphi} + \frac{\partial \eta}{\partial \lambda} = U \cos\varphi \frac{\partial (\mathfrak{z} + 2\tau)}{\partial U} + \frac{2 \rho \omega^2}{\gamma} (\mathfrak{z} + 2\tau) \cos\varphi \,. \tag{20}$$

Ignoring the last term, and noting that $\mathfrak{z} + 2\tau = \frac{\partial T}{\partial U}$, we also have

$$\frac{\partial \xi \cos\varphi}{\partial \varphi} + \frac{\partial \eta}{\partial \lambda} = U \cos\varphi \frac{\partial^2 T}{\partial U^2} \,. \tag{20'}$$

Bibliography

Heiskanen W, Moritz H (1967) Physical geodesy. Freeman, San Francisco

Hirvonen RA (1960) New theory of the gravimetric geodesy. Isost Inst IAG 32

Hirvonen RA (1961) The reformation of geodesy. J Geophys Res 66 (5)

Hotine M (1969) Mathematical geodesy. US Dep Commerce, ESSA Monogr 2

Marussi A (1950a) Sviluppi di Legendre generalizzati per una curva qualunque tracciata su di una superficie pure qualunque. Rend Accad Naz Lincei 8: Ser 8 (4) (Chap. I.3, this Vol.)

Marussi A (1950b) Sviluppi di legendre generalizzati nello spazio. Rend Accad Naz Lincei 8: Ser 8 (4) (Chap. I.4, this Vol.)

Marussi A (1950c) Sviluppi di Legendre generalizzati per una curva qualunque dello spazio. Rend Accad Naz Lincei 9: Ser 8 (1 – 2) (Chap. I.5, this Vol.)

Marussi A (1950d) Principi di geodesia intrinseca applicati al campo di Somigliana. Boll Geodes Sci Affini 8 (2) (Chap. III.1, this Vol.)

Marussi A (1951a) Les principes de la géodésie intrinsèque. Bull Géodés Assoc Int Géodésie 19

Marussi A (1951b) Fondamenti di Geodesia intrinseca. Pubbl Comm Geodet Ital Terza Ser Mem 7 (Chap. I.2, this Vol.)

Marussi A (1953) Sulla divergenza del campo gravimetrico normale. Geofis Pura Appl 25

Marussi A (1957) Sulle rappresentazioni fra superfici definite mediante la forma quadratica che ne determina il modulo di deformazione. Festschr CF Baeschlin, Zürich (Chap. V.3, this Vol.)

Marussi A (1959) Dalla geodesia classica alla geodesia in tre dimensioni. Boll Geodes Sci Affini 18 (4) (Chap. I.1, this Vol.)

Marussi A (1961) Intrinsic coordinates in practical geodesy. The Earth today. Geophys J R Astron Soc. Vol. 4

Marussi A (1973) On the representation of the actual gravity field of the Earth on the normal ellipsoidal field. Geophys J R Astron Soc London 37:3

4 The Adjustment of Geodetic Networks in the Three-Dimensional Ellipsoidal Model Space of Somigliana [1]

1. The ultimate aim in the computation of geodetic networks is the establishment of a one-to-one correspondence between points of the natural geodetic field and sets of coordinates associated with a metric tensor, enabling the recovery of metric relations between such points. The coordinates are furthermore required to be close to observable parameters in the natural geodetic field and the associated metric tensor to be of simple analytical structure. A scale factor may be assigned to account for possible strains induced by the procedure in the metric relations.

The procedure may be alternatively described as the mapping of the actual natural geodetic field onto a mathematical model field endowed with appropriate coordinates; commonly the gravitational and rotational potential field based on a best-fitting ellipsoid, endowed with latitude, longitude, and geopotential as coordinates.

2. In the classical procedure followed for over two centuries, the mapping has been confined to the two-dimensional domain, the objective space being the physical surface of the Earth and the image space the surface of a best-fitting reference ellipsoid. In this procedure, the mapping takes place in two distinct phases: the projection of the physical surface onto the geoid, and the projection (or mapping) of the geoid onto the ellipsoid.

The first phase is performed by the well-known series of "reductions" taking into account the height a.s.l. (or above the geoid) of the individual points; the height is thus eliminated at the first step. The second phase consists of the projection (or mapping) of the geoid onto the ellipsoid, taking into account the separation between the two surfaces. For physical reasons, it is assumed that both the geoid and the reference ellipsoid are characterized by the same value of the potential.

Having thus treated separately the height, the remaining two coordinates are the ellipsoidal or geodetic latitude φ, and the ellipsoidal or geodetic longitude λ. These are computed "a posteriori" from the metric quantities actually measured on the field and reduced according to the described procedure, via the metric tensor defined by the quadratic form

$$ds^2 = \rho^2 d\varphi^2 + \nu^2 \cos^2 \varphi d\lambda^2 \tag{1}$$

[1] Originally published as: Marussi A (1981) The adjustment of geodetic networks in the three-dimensional ellipsoidal model space of Somigliana. Proc Symp Geodet Networks Comput, Munich. DGK, B258/VI

in which ρ and ν are the radii of curvature of the meridian and the grand normal of the reference ellipsoid, respectively. The algorithm used is based on the expansions of Legendre and Delambre.

The classical procedure has obvious drawbacks, both conceptual and practical:

Conceptual, inasmuch as it discriminates between the three coordinates, because it refers to the geoid, an ill-defined and inaccessible surface under the continents, and finally since it is based on the consideration of fictitious geometrical objects, surface geodesics, that have no physical reality.

Practical, inasmuch as the reduction to the geoid affects at the outset all observed quantities, lengths and angles, by drastically altering them with a series of ill-defined numerical operations that downgrade by orders of magnitude the high accuracies that can be obtained with modern instruments. Conversely, inasmuch as the metric quantities recoverable via the metric tensor have no immediate counterparts in physical quantities measurable in the actual field.

3. Space techniques have radically changed the scene of Geodesy in recent times, and the vertical dimension has now entered in that scene at equal rights with the horizontal dimensions. It therefore appears that the classical treatment that discriminates between the three coordinates needs a thorough revision.

The present chapter aims at sketching a practical procedure based on the rigorous principles of three-dimensional mapping, that generalizes in the most immediate way the classical procedure. The generalization is reached by introducing the model potential field U based on the conventional rotating reference ellipsoid and requiring that now not only the geoid, but any other equipotential surface of the actual potential field W, be mapped onto the level surface of the model field U characterized by the same value of the potential. Any point P of the actual geodetic field located on the equipotential surface of potential W will consequently be mapped on a point Q of the model field U such that $U(Q) = W(P)$.

In order to complete the correspondence, the model field is superposed onto the actual field in such way that its centre of figure coincides with the centre of mass and its axis of figure coincides with the mean axis of rotation of the Earth. Space techniques make the fulfilment of these requirements possible.

Indicating again with φ and λ the geodetic latitude and longitude in the model field, the correspondence is then completed by putting

$$\varphi(Q) = \varphi(P)\,; \qquad \lambda(Q) = \lambda(P)\,; \qquad U(Q) = W(P)\,. \tag{2}$$

From the geometrical point of view, the procedure is equivalent to the projection of P onto Q along the isozenith line of the model field. Such procedure generalizes the one followed in the conventional computation of networks, and keeps the induced metric distorsions to a minimum.

4. It has been shown elsewhere (Marussi 1973, 1974a, 1974b) that the metric tensor of the normal field is given by

$$ds^2 = \rho^2 d\varphi^2 + \nu^2 \cos^2\varphi\, d\lambda^2 + \frac{2\rho f}{\gamma}\, d\varphi\, d\lambda + \frac{1+f^2}{\gamma^2}\, dU^2 \tag{3}$$

or, introducing the dynamical height h defined by

$$h = -\frac{U - U_0}{\gamma_0} \tag{4}$$

in which U_0 is the potential on the base ellipsoid, by the equivalent form

$$ds^2 = \rho^2 d\varphi^2 + \nu^2 \cos^2\varphi d\lambda^2 - 2\rho q f d\varphi dh + q^2(1+f^2) dh^2 \tag{5}$$

where ρ and ν are now the radius of curvature and the grand normal of the equipotential surface U at Q, γ the normal gravity at Q, $f = \frac{\partial \lg \gamma}{\partial \varphi}$, and $q = \frac{\gamma_0}{\gamma}$. The values of ρ, ν and γ are easily computed starting from the values assigned on the base ellipsoid remembering that (Marussi 1950c)

$$\frac{\partial \rho}{\partial U} = -\frac{1}{\gamma}\left(1 + f^2 - \frac{\partial f}{\partial \varphi}\right)$$

$$\frac{\partial \nu}{\partial U} = -\frac{1}{\gamma}(1 + f \operatorname{tg}\varphi)$$

$$\frac{\partial \gamma}{\partial U} = \frac{1}{\nu} + \frac{1+f^2}{\rho} + \frac{2\omega^2}{\gamma} \tag{6}$$

in which ω is the angular speed of rotation of the Earth. γ_0 is a conventional value of γ used for the conversion of potentials into dynamical heights.

5. It has further been shown elsewhere (Marussi 1973, 1974b) that the strain induced by the $P \rightarrow Q$ mapping is given by

$$m = \frac{ds_Q}{ds_P} = 1 - \tau \sin^2 z + 3 \cos^2 z + \frac{1}{2}(\xi \cos\alpha + \eta \sin\alpha) \sin 2z \tag{7}$$

in which z is the zenith distance of the displacement considered,

$$\tau = \frac{T}{U}, \qquad \mathfrak{z} = \frac{\Delta g}{\gamma}, \tag{8}$$

$$T(P) = W(P) - U(P), \qquad \Delta g(P) = g(P) - \gamma(Q) \tag{9}$$

and ξ and η are the deflections of the vertical

$$\xi = \Phi(P) - \varphi(P)$$

$$\eta = (\Lambda(P) - \lambda(P)) \cos\varphi \tag{10}$$

Φ and Λ being the (astronomical) latitude and longitude in the W field, respectively.

6. The correspondence between objective points P of the natural field and points Q of the model field having thus been established along with its metrical consequences, the mapping of actual geometrical objects onto the model space will require the reduction of all measured quantities (distances and angles) according to the strain tensor on which Eq. (7) is based.

It should be recalled that in the correspondence the geodetic meridians of the actual field (not the North lines) are mapped onto the meridians of the model field; previous to mapping, astronomically measured azimuths should be reduced for the azimuth of the geodetic meridian (Marussi 1951, p. 20; this Vol., p. 30).

Having thus scaled and oriented in the model field the geometrical objects observed in the actual field, the next problem will be the computation in an assigned datum of the coordinates φ and λ and h of the vertices of such objects. It is well known that this problem goes under the name of first fundamental problem in geometrical Geodesy.

The classical solution of the problem is given by the expansions of Legendre and Delambre generalized in the three-dimensional space (Marussi 1950a, b) and specialized for the ellipsoidal or Somigliana field (Marussi 1950d).

As we shall see, the solution of the inverse problem (second fundamental problem), i.e., the computation of lengths, azimuths and zenith distances from coordinates, will also be required for the application of the method of adjustment by variation of coordinates. The solution can be easily obtained by inverting Legendre's expansions by means of a procedure of successive approximations.

7. Having thus outlined the theoretical principles of the proposed three-dimensional mapping, the actual computation of a geodetic network can be easily performed by the method of least squares applying the technique of variation of coordinates. The great advantage of this technique is that no actual adjustment in three-dimensional space is required, and that the coordinates determined by the conventional procedures can be used as approximate coordinates.

The only operation to be performed in the three-dimensional space will in fact be the rigorous computation by means of the inverted Legendre's developments of the lengths and angles from the assigned approximate coordinates that need to be compared with the observed ones after the due reduction for the strain induced by the mapping.

8. It should be further noted that:

a) Owing to the very weak correlation that exists between height and geographical coordinates, it will in general be possible to split the adjustment into two steps: first, the adjustment of heights considering both spirit levelling and trigonometrical measurements; second, the adjustment of the geographical coordinates. An iterative procedure may be applied, if necessary.

b) The quantity f appearing in formulae (8) and (5) that characterize the metric tensor of the mathematical model field, never exceeds $5 \cdot 3^{-3}$ and the quantity q differs from unity by at most $2 \cdot 8^{-3}$. Consequently, the rectangular term will be generally negligible, and the coefficient of dh^2 will differ from unity by not more than 6^{-3}.

c) The quantities τ, $\mathfrak{z}$, ξ, and η that appear in formula (7) that characterizes the strain tensor, should be computed from an expansion in terms of spherical harmonics of a realistic model of the gravity field derived from satellite and surface data. The values of the quantities referred to may be tabulated for easy reference, represented in graphical form, or programmed.

Bibliography

Hirvonen RA (1960) New theory of the gravimetric geodesy. Isost Inst IAG no 32

Hirvonen RA (1961) The reformation of geodesy. J Geophys Res 66 (5)

Marussi A (1950a) Sviluppi di Legendre generalizzati nello spazio. Rend Accad Naz Lincei 8: Ser 8 (5) (Chap. I.4, this Vol.)

Marussi A (1950b) Sviluppi di Legendre generalizzati per una curva qualunque dello spazio. Rend Accad Naz Lincei 9: Ser 8 (1 – 2) (Chap. I.5, this Vol.)

Marussi A (1950c) Principi di Geodesia intrinseca applicati al campo di Somigliana. Boll Geodes Sci Affini 8 (2) (Chap. III.1, this Vol.)

Marussi A (1950d) Il primo problema fondamentale della Geodesia ampliato nel campo di Somigliana. Geofis Pura Appl 18 (Chap. III.2, this Vol.)

Marussi A (1951) Fondamenti di Geodesia intrinseca. Comm Geodet Ital Mem 7 (Chap. I.2, this Vol.)

Marussi A (1961) Intrinsic coordinates in practical geodesy. Geophys J Roy Astron Soc 4

Marussi A (1973) On the three-dimensional computation of geodetic networks and anomalies. Rend Accad Naz Lincei 55: Ser 8 (5) (Chap. IV.3, this Vol.)

Marussi A (1974a) On the representation of the actual gravity field of the Earth on the normal ellipsoidal field. Geophys J Roy Astron Soc 37:3

Marussi A (1974b) Geodetic networks in space. Can Surv 28 (5)

V Mapping Between Surfaces

1 Some Integral Properties of the Conformal Representations of Surfaces on Surfaces[1]

1. The well-known theorem of Gauss-Bonnet (Bonnet 1848) for a regular surface S expresses the integral tangential curvature along a closed circuit c free of nodes, but with a discrete number of angular points, in terms of the integral curvature of the area A contained within c, and of the sum of the angles formed by the tangents to the curve at its angular points; it is a generalization of the theorem of Gauss which furnishes the angular excess of a geodesic triangle inscribed on any surface in terms of its integral curvature, and which reduces to the theorem of Cavalieri in the case of the spherical triangle.

Let us now represent S conformally on another surface S', so the circuit c will be transformed into the circuit c'. Since the tangential curvature c' is linked to that of c by Schols' theorem, which introduces, as well as the modulus of deformation, its first derivatives also, then from a comparison between the two expressions for the integral tangential curvature one obtains straight away that the change in this curvature equals the total flux of the *gradient of deformation* across the circuit, or also, by Green's formula, the surface integral on A of the divergence of this gradient. Thus the Gauss-Bonnet theorem applied in turn to the two surfaces S and S' allows us to recover by other means the differential equation that links the modulus of deformation to the total curvatures of the two surfaces.

2. On the given surfaces S and S', both regular, let us define respectively

$$\left.\begin{aligned} ds^2 &= a_{ij}dx^i dx^j \\ ds'^2 &= a'_{ij}dx^i dx^j \end{aligned}\right\} (i, j = 1, 2)\,. \tag{1}$$

as the expressions for the square of the linear element, in which it is understood, following the conventions of the tensor calculus, that summation is performed with respect to the indices which appear twice in each term, once as superscript and once as subscript.

Let us suppose that the two surfaces are in a conformal representation, and referred to the same coordinates x^i, we shall have

$$a'_{ij} = e^{2\mu} a_{ij}\,; \qquad a'^{\,ij} = e^{-2\mu} a^{ij} \tag{2}$$

in which μ, a function of the x^i only (and not of their derivatives), is the natural logarithm of the modulus of deformation, and a^{ij} is the tensor a_{ij} in contravariant form.

[1] Originally published as: Marussi A (1951) Su alcune proprietà integrali delle rappresentazioni conformi di superfici su superfici. Rend Accad Naz Lincei 10: Ser 8, fasc 4

We shall similarly call K and K' the total curvatures of the two surfaces at points which correspond in the chosen representation.

3. Given the above, we shall consider on S a closed curve c, free of nodes, which bounds the regular part A of the surface within it; we shall not exclude that the curve c may have a finite number n of angular points. Denoting in this latter case by θ_α $(\alpha = 1,2,\ldots,n)$ the external angles formed by the normals at the angular points, and by γ the tangential curvature of c (positive when the curve is convex towards the interior), the Gauss-Bonnet theorem (Eisenhart 1947) states

$$\iint_{(A)} K dS = 2\pi - \sum_{\alpha=1}^{n} \theta_\alpha + \int_{(c)} \gamma ds = 2\pi - \sum_{\alpha=1}^{n} \theta_\alpha - \Gamma \tag{3}$$

in which the first surface integral represents the integral curvature of the area A, Γ the integral tangential curvature of the circuit, traversed leaving the area A to the right, and $dS = \sqrt{\Delta}\, dx^1 dx^2$ the element of area, Δ being the determinant of the a_{ij}.

If now μ is in turn regular within the area considered, a completely analogous formula holds on S' for the curve c', the transform of c in the chosen conformal representation. We note that from the conformality of the representation $\theta'_\alpha = \theta_\alpha$, and that by Schols' theorem which provides the change in tangential curvature we have

$$\gamma' = e^{-\mu}\gamma - \frac{d\mu}{dn} \tag{4}$$

where by γ' is denoted the tangential curvature of c'; $d\mu/dn$ represents then the derivative of μ taken at the points of c' in the direction of the normal $\boldsymbol{n}'$ to the curve itself, with $\boldsymbol{n}'$ such that the pair of versors $(\boldsymbol{n}', \boldsymbol{t}')$ are right-handed ($\boldsymbol{t}'$ is the versor of the tangent). If ν^i are the contravariant components of $\boldsymbol{n}'$ we have then

$$\frac{d\mu}{dn} = \operatorname{grad}\mu \times \boldsymbol{n}' = \frac{\partial \mu}{\partial x^i} \nu^i \tag{5}$$

and taking note that $ds' = e^{\mu} ds$ we have also $(\mu_i = \partial\mu/\partial x^i)$

$$\int_{(c')} \gamma' ds' = \int_{(c)} \gamma ds - \int_{(c')} \mu_i \nu^i ds \tag{6}$$

and therefore

$$\Gamma' = \Gamma - \int_{(c')} \operatorname{grad}\mu \times \boldsymbol{n}' ds' = \Gamma - \int_{(c')} \mu_i \nu^i ds' . \tag{7}$$

For simplicity we shall call the vector $\operatorname{grad}\mu$ the gradient of deformation, whose covariant components are μ_i on both S and S'; and we shall have then that the change in the integral tangential curvature of a closed circuit equals, in a conformal representation, the flux of the gradient of deformation across the circuit itself.

From the invariant character of the operations considered, and from the conformality of the representation, it is the same whether this flux is calculated on S or on S'.

4. We may note now that, under the hypothesis that μ is regular within the area considered, the line integral that provides the above flux can be transformed, with Green's formula, into a surface integral extending over the whole area enclosed by the circuit, according to the formula (Eisenhart 1947, Burgatti 1951)

$$\int_{(c')} \mu_i v^i ds' = \iint_{(A')} \mu'^i/_i dS' \tag{8}$$

in which $\mu'^i/_i$ represents the second differential parameter $(\Delta_2\mu)_{S'} = \operatorname{div}\operatorname{grad}_{S'}\mu$, of μ on S'. From this, denoting by Δ' the determinant of a'_{ij} we have

$$\Delta' = e^{4\mu}\Delta\,; \qquad a'^{ij} = e^{-2\mu}a^{ij} \tag{9}$$

and hence

$$\mu'^i/_i = \frac{1}{\sqrt{\Delta'}}\,\frac{\partial(\sqrt{\Delta'}\,a'^{ij}\mu_j)}{\partial x^i} = e^{-2\mu}\frac{1}{\sqrt{\Delta}}\,\frac{\partial(\sqrt{\Delta}\,a^{ij}\mu_j)}{\partial x^i} = e^{-2\mu}\mu^i/_i\,, \tag{10}$$

we can then write also

$$\Gamma' - \Gamma = -\iint_{(A')} (\Delta_2\mu)_{S'} dS' = -\iint_{(A)} (\Delta_2\mu)_S\, dS \tag{11}$$

because obviously $dS' = e^{2\mu}dS$.

We have then: the change in the integral tangential curvature of a closed circuit likewise equals, in a conformal representation, the surface integral of the divergence of the gradient of deformation extended over the internal area of the circuit.

5. From the above and from Eq. (3) it follows that

$$\begin{aligned} \Gamma' - \Gamma &= -\iint_{(A)} K\,dS + \iint_{(A)} e^{2\mu}K'\,dS = -\iint_{(A)} (\Delta_2\mu)_S\, dS \\ &= -\iint_{(A')} e^{-2\mu}K\,dS' + \iint_{(A')} K'\,dS' = -\iint_{(A')} (\Delta_2\mu)_{S'} dS' \end{aligned} \tag{12}$$

and hence, from the arbitrary choice of the circuit considered,

$$\left.\begin{aligned} \mu^i/_i &= (\Delta_2\mu)_S = K - K'e^{2\mu} \\ \mu'^i/_i &= (\Delta_2\mu)_{S'} = Ke^{-2\mu} - K' \end{aligned}\right\}. \tag{13}$$

These formulae had already been found by another way (Marussi 1951), and are now put in a new light by the considerations expounded here, which provide a relationship between Schols' theorem and the differential equation which connects the modulus of deformation with the total curvatures of the two surfaces in a conformal representation.

6. It follows as a corollary that when the surface S' is plane (theory of geographic maps), and thus $K' = 0$, the change in the integral tangential curvature of a circuit equals the integral curvature of the object area enclosed by it on S, and thus the angular-excess above 2π on its boundary; an excess which can similarly be measured by the total flux of the gradient of deformation across the circuit.

In the even more particular case when the cirucit c on S is a geodesic triangle ($\Gamma = 0$), we have then for the angular excess

$$2\pi - \sum_{\alpha=1}^{3} \theta_\alpha = \int\int_{(A')} (\Delta_2 \mu)_{S'} dS' = \int_{(c')} \operatorname{grad} \mu \times \boldsymbol{n}' ds' \tag{14}$$

i.e., the angular excess of a geodesic triangle equals the total flux of the gradient of deformation of its image in any conformal representation whatever on the plane, or alternatively, the integral of the divergence of this gradient, extending over the whole internal area of the said image.

If then both surfaces S and S' were plane, we would have

$$\Gamma = \Gamma' \tag{15}$$

and hence: in any conformal representation whatever of the plane on itself, the change in the integral tangential curvature is zero for any circuit; and this is an immediate consequence of the harmonicity, in this case, of the logarithm of the modulus of deformation.

Bibliography

Bonnet O (1848) Mémoire sur la théorie générale des surfaces. J Ecole Polytech 32:131

Burgatti P (1951) I teoremi del gradiente, della divergenza, della rotazione sopra una superficie, ecc. R Accad Sci Ist Bologna 4: Ser 7

Eisenhart LP (1947) An introduction to differential geometry. Princeton Univ Press, Princeton, p 191

Marussi A (1951) Su alcune proprietà fondamentali delle rappresentazioni conformi fra superfici qualunque. Boll Geodes no 2

2 The a Priori Determination of the Modulus of Linear Deformation in the Gauss Conformal Representation[1]

1. The ever wider development of the conformal representation of Gauss, or Transverse Mercator, which has also recently been introduced in Italy and is widely used under the name of the Gauss-Boaga representation, has brought into being a vast number of studies, some with purely theoretical ends, others with a practical aim (Marussi 1950). Because of this, it has seemed to me not inappropriate to make a contribution to the knowledge of this most important representation, by showing how one can obtain, from the logarithm of its modulus of deformation, a particularly simple and expressive series expansion, whose deduction does not require previous knowledge of the equations of the representation itself. I think it is not without purpose to draw attention to these expansions, since they also lend themselves to extension to more complex cases.

2. In two other works (Marussi 1951a; Chap. V.1, this Vol.) we have seen how in any conformal representation of any surface σ whatever on any other surface σ' whatever, the natural logarithm μ of the modulus of linear deformation should satisfy the equations in partial derivatives (which are mutually equivalent)

$$(\Delta_2\mu)_\sigma = K - e^{2\mu}K' \tag{1'}$$

$$(\Delta_2\mu)_{\sigma'} = e^{-2\mu}K - K' \tag{1''}$$

in which K and K' are, respectively, the total curvatures of σ and of σ' at points which correspond in the given representation, and $(\Delta_2 u)_\sigma$ is the second differential parameter of Beltrami calculated on σ, $(\Delta_2\mu)_{\sigma'}$ is the same quantity calculated on σ'. The equations written here represent the equation of Laborde in absolute form.

When σ' is a plane, $K' = 0$, and the above equations are simplified as

$$(\Delta_2\mu)_\sigma = K \qquad (\sigma' \text{ a plane}) \tag{2'}$$

$$(\Delta_2\mu)_{\sigma'} = e^{-2\mu}K; \tag{2''}$$

the first being thus equivalent to an equation of Poisson's type.

We now propose to integrate these equations, for the case in which σ is a rotational ellipsoid (or more generally, any regular surface of rotation), supposing that we can refer this surface to a system of intrinsic geographic coordinates, and under the hypothesis that the given representation shall be that of Gauss.

---

[1] Originally published as: Marussi A (1951) Determinazione a priori del modulo di deformazione lineare nella rappresentazione conforme di Gauss. Rend Accad Naz Lincei 11: Ser 8, fasc 3–4

3. If a_{ij} is the fundamental tensor of σ in the chosen coordinates x^i, and a^{ij} the associated (reciprocal) tensor, we may write for the general expression of $(\Delta_2\mu)_\sigma$

$$(\Delta_2\mu)_\sigma = \frac{1}{\sqrt{A}} \frac{\partial(\sqrt{A}\, a^{ij}\, \partial\mu/\partial x^j)}{\partial x^i} \tag{3}$$

where A is the determinant of the a_{ij}. In Eq. (3) it is obviously intended that summation should be carried out for the values 1 and 2 of all indices, such as i, j, which appear twice.

Equation (3), however, is simplified if, as in our case, the system of coordinates x^i is orthogonal, whence $a_{12} = a^{12} = 0$. Setting then

$$a_{11} = H_1^2; \qquad a_{22} = H_2^2 \tag{4}$$

we have

$$a^{11} = \frac{1}{H_1^2}; \qquad a^{22} = \frac{1}{H_2^2}; \qquad A = H_1^2 H_2^2 \tag{4'}$$

and (3) becomes

$$(\Delta_2\mu)_\sigma = \frac{1}{H_1 H_2}\left[\frac{\partial}{\partial x^1}\left(\frac{H_2}{H_1}\frac{\partial\mu}{\partial x^1}\right) + \frac{\partial}{\partial x^2}\left(\frac{H_1}{H_2}\frac{\partial\mu}{\partial x^2}\right)\right]. \tag{3'}$$

If the system of coordinates is moreover isothermal, and introducing the isometric parameters, Eq. (3′) is further simplified, since under this assumption $H_1 = H_2 = H$, and hence

$$(\Delta_2\mu)_\sigma = \frac{1}{H^2}\left(\frac{\partial^2\mu}{\partial x^{1^2}} + \frac{\partial^2\mu}{\partial x^{2^2}}\right). \tag{3''}$$

4. In the case of a surface of rotation referred to intrinsic geographic coordinates we have (φ latitude, λ longitude)

$$ds^2 = \rho^2 d\varphi^2 + N^2\cos^2\varphi\, d\lambda^2 \tag{5}$$

with ρ the radius of curvature of the meridian, N of the prime vertical; or else, introducing the isothermal-isometric latitude Φ

$$d\Phi = \frac{\rho\, d\varphi}{N\cos\varphi} \tag{6}$$

$$ds^2 = N^2\cos^2\varphi(d\Phi^2 + d\lambda^2); \tag{5'}$$

and hence, in this case

$$(\Delta_2\mu)_\sigma = \frac{1}{N^2\cos^2\varphi}\left(\frac{\partial^2\mu}{\partial\Phi^2} + \frac{\partial^2\mu}{\partial\lambda^2}\right). \tag{7}$$

Equation (2′) will thus be written in these coordinates, recalling that $K = 1/\rho N$,

$$\frac{\partial^2\mu}{\partial\Phi^2} + \frac{\partial^2\mu}{\partial\lambda^2} = \frac{N\cos^2\varphi}{\rho} = \mathcal{R}. \tag{8}$$

This is a partial differential equation of the second order of elliptical type (of Poisson). Its integration depends on a knowledge of μ and of its normal derivative $d\mu/dn$ along an arbitrarily given curve γ on σ (the curve carrying the boundary values).

In the case of the Gaussian representation, we select as such a curve on the ellipsoid (or surface of rotation) a meridian, and we assume that on this curve $\mu = d\mu/dn = 0$. Let us see how, with these conditions, it is quite easy to determine μ by means of series integration.

5. To do this, we shall assume for μ the following form of series expansion in powers of λ

$$\mu = \mu_0 + \mu_1 \lambda + \mu_2 \lambda^2 + \mu_3 \lambda^3 + \dots \tag{9}$$

justified by the consideration that, since the Gaussian representation is, as we know, limited to a rather narrow meridional zone, it is convenient to adopt an expansion in powers of λ which assures a rapid convergence of the series. In this series, the μ_r are functions of latitude φ only (or, what is the same thing, of the isometric latitude Φ alone). We have immediately

$$\left.\begin{aligned} \frac{\partial\mu}{\partial\lambda} &= \mu_1 + 2\mu_2\lambda + 3\mu_3\lambda^2 + \cdots + r\mu_r\lambda^{r-1} + \dots \\ \frac{\partial^2\mu}{\partial\lambda^2} &= 2\mu_2 + 6\mu_3\lambda + 12\mu_4\lambda^2 + \cdots + r(r-1)\mu_r\lambda^{r-2} + \dots \\ \frac{\partial^2\mu}{\partial\Phi^2} &= \mu_0'' + \mu_1''\lambda + \mu_2''\lambda^2 + \mu_3''\lambda^3 + \dots \end{aligned}\right\} \tag{10}$$

having set $\mu_r'' = \partial^2\mu_r/\partial\Phi^2$.

On the curve γ, whose equation we shall assume is given by $\lambda = 0$, we should have

$$\mu = 0\,; \qquad \frac{d\mu}{dn} = 0 \qquad (\text{on } \gamma) \tag{11}$$

whence it follows immediately that

$$\mu_0 = \mu_1 = 0\,; \tag{12}$$

then, substituting in Eq. (8) the values found for the second derivatives

$$\mu_2''\lambda^2 + \mu_3''\lambda^3 + \cdots + 2\mu_2 + 6\mu_3\lambda + 12\mu_4\lambda^2 + \cdots = \mathscr{R}\,; \tag{8$'$}$$

whence, collecting terms in the same powers of λ

$$\begin{aligned} 2\mu_2 - \mathscr{R} + 6\mu_3\lambda + (\mu_2'' + 12\mu_4)\lambda^2 + (\mu_3'' + 20\mu_5)\lambda^3 + (\mu_4'' + 30\mu_6)\lambda^4 + \dots \\ + [\mu_r'' + (r+1)(r+2)\mu_{r+2}]\lambda^r + \cdots = 0\,. \end{aligned} \tag{8$''$}$$

Since the series is identically zero, we should then have

$$
\left.
\begin{aligned}
&\mu_3 = \mu_5 = \cdots = \mu_{2r+1} = \cdots = 0 \\
&\mu_2 = \frac{1}{2!}\,\mathscr{R}, \\
&\mu_4 = -\frac{\mu_2''}{12} = -\frac{1}{4!}\,\frac{d^2\mathscr{R}}{d\Phi^2} \\
&\mu_6 = -\frac{\mu_4''}{30} = \frac{1}{6!}\,\frac{d^4\mathscr{R}}{d\Phi^4} \\
&\cdots\cdots\cdots\cdots\cdots \\
&\mu_{2r} = -\frac{\mu_2''{}_{(r-1)}}{4r(r-1)} = \frac{(-1)^{r-1}}{(2r)!}\,\frac{d^{2(r-1)}\mathscr{R}}{d\Phi^{2(r-1)}};
\end{aligned}
\right\} \tag{13}
$$

the calculation of the derivatives with respect to Φ should be straightforward if we remember that formally

$$
\frac{d}{d\Phi} = \frac{d\varphi}{d\Phi}\,\frac{d}{d\varphi} = \frac{N\cos\varphi}{\rho}\,\frac{d}{d\varphi}. \tag{14}
$$

6. The logarithm μ of the modulus of deformation thus comes to be expressed, in the Gaussian representation of any surface of rotation on the plane, by the rather simple series expansion

$$
\begin{aligned}
\mu = {} & \frac{\lambda^2}{2!}\,\mathscr{R} - \frac{\lambda^4}{4!}\,\frac{d^2\mathscr{R}}{d\Phi^2} + \frac{\lambda^6}{6!}\,\frac{d^2\mathscr{R}}{d\Phi^4} + \ldots \\
& + (-1)^{r-1}\,\frac{\lambda^{2r}}{(2r)!}\,\frac{d^{2(r-1)}\mathscr{R}}{d\Phi^{2(r-1)}} + \ldots .
\end{aligned} \tag{15}
$$

Bibliography

Marussi A (1950) La teoria della proiezione di Gauss con premesse di geometria differenziale. Lecture held at 12th Corso di Geodesia dell' Ist. Geogr Militare, Florence, 1950

Marussi A (1951a) Su alcune proprietà fondamentali delle rappresentazioni conformi fra superfici. Boll Geodes no 1

3 Representations Between Surfaces Defined by Means of the Quadratic Form Which Determines Their Modulus of Deformation[1]

1. In earlier works I have had occasion to study representations between surfaces, in which it is assumed that, on one of the surfaces (which we shall call the object surface) there is arbitrarily assigned the quadratic form as a function of position or the surface, which fixes in every direction the modulus of linear deformation of the representation itself. If the quadratic form is quite general, limited only in that it should be positive definite, and if its coefficients are continuous functions together with their first and second derivatives, the representation will be said to be affine, since it establishes an affinity between the neighbourhood of the first order of the object surface Σ, and the corresponding neighbourhood of the second surface $\bar{\Sigma}$ (which we shall call the image surface).

In these works I have dwelt in particular upon conformal representations, characterized by an "isotropic" quadratic form, which in turn can be expressed as a scalar function of position (the modulus of deformation, this time independent of direction); then I was concerned with the law according to which the tangential curvature of any given curve is transformed in a general affine transformation.

In the present chapter, which I offer in respectful homage to Professor C. F. Baeschlin, to whom geodetic study owes so much, on the occasion of this 75th birthday, I have sought to take up again a part of these investigations, casting them in a synthetic mould, and then finding the relationship between the total curvatures of the two surfaces in an affine representation, expressed by means of the ordinary and differential invariants of the above-mentioned quadratic form.

Particular applications are then made in the case of conformal, geodetic, and equivalent representations, finding for these several general properties, in part already known, and in part however new.

The study which I have briefly sketched was carried out by making use of the methods of the tensor calculus, which offer the particular advantage of immediately showing up the invariant properties of the geometric entities which have a significance independent of the coordinates used, and yet provide the components of these entities as required in practical calculation at the moment of application. These methods further allow us to give, as far as possible, a synthetic and closed form to the formulas, otherwise rather complex, and are apt for ready generalisations to spaces of more than two dimensions, in accord with the directions in which geodetic studies are now progressing.

---

[1] Originally published as: Marussi A (1957) Sulle rappresentazioni fra superfici definite mediante la forma quadratica che ne determina il modulo di deformazione. Festschr CF Baeschlin, Zürich

From this viewpoint, the present investigations are related to those made with different objectives, and to cite only a few names, by T. Levi-Civita, by A. Finzi, by A. S. Eddington, and by T. A. Schouten, on spaces with affine or conformal connexions, spaces which, for the varieties with two dimensions, give a concrete geometric image.

2. Let us then take a regular part Σ of a surface, endowed with a double system of coordinate lines so as to fix a reciprocal correspondence between the pairs of coordinates which we shall call x^1 and x^2, and the points of Σ; and let us also take a positive definite quadratic form

$$\sum_{i,j=1,2} g_{ij}\, dx^i dx^j = ds^2 \qquad (g_{ij} = g_{ji}) \tag{1}$$

in the differentials of the coordinates, which yields the square of the modulus of the displacement of components dx^1 and dx^2. Dividing Eq. (1) by ds^2, and setting

$$\lambda^1 = \frac{dx^1}{ds}; \qquad \lambda^2 = \frac{dx^2}{ds}; \qquad \lambda^i = \frac{dx^i}{ds} \tag{2}$$

it follows that

$$g_{ij}\lambda^i\lambda^j = 1\,. \tag{3}$$

Here, as is usual in the tensor calculus, whose methods we shall follow from now on, we have suppressed the symbol of summation with respect to the indices which appear twice in the formulae, once above and once below, indices which without exception can take the values 1 and 2. It turns out that the λ^i can be considered as the components of a unit vector (or versor) parallel to the displacement of components dx^i. It is well known that the dx^i, or the λ^i, are transformed by contravariance, while the coefficients g_{ij} of Eq. (3) are transformed by covariance; the latter are thus the components of a double covariant tensor which will be called, because of its significance, the fundamental metric tensor on Σ (with respect to the coordinates x^i).

Let us assume that the g_{ij} are continuous together with their first and second derivatives, and that from the fundamental metric tensor we shall construct the Christoffel symbols of the first kind on Σ:

$$(ij, h) = (ji, h) = \frac{1}{2}\left(\frac{\partial g_{hi}}{\partial x^j} + \frac{\partial g_{jh}}{\partial x^i} - \frac{\partial g_{ij}}{\partial x^h}\right) \tag{4}$$

and those of the second kind

$$\begin{Bmatrix} k \\ i \;\; j \end{Bmatrix} = \begin{Bmatrix} k \\ j \;\; i \end{Bmatrix} = g^{hk}(ij, h) \tag{5}$$

where g^{hk} are the reciprocal elements of the matrix of the tensor g_{ij},

$$g_{ij}g^{jh} = \delta_i^h = \begin{cases} 0 & i \neq h \\ 1 & i = h \end{cases} \tag{6}$$

in which δ_i^h represents Kronecker's tensor.

Let us put also

$$G = | g_{ij} |; \qquad \frac{1}{G} = | g^{ij} | \tag{7}$$

and recall that

$$\left\{ \begin{matrix} j \\ i \quad j \end{matrix} \right\} = \left\{ \begin{matrix} j \\ j \quad i \end{matrix} \right\} = \frac{\partial}{\partial x^i} \lg \sqrt{G}\,. \tag{8}$$

3. Let us now take on Σ a second positive definite quadratic form

$$a_{ij} \lambda^i \lambda^j = m_\lambda^2 \qquad (a_{ij} = a_{ji}) \tag{9}$$

$$a_{ij} dx^i dx^j = m_\lambda^2 ds^2 = \overline{ds^2} \tag{9'}$$

whose coefficients we shall take to be continuous functions of the coordinates (x^i) along with, at least, their first and second derivatives.

This form determines, for every direction λ^i issuing from the point of coordinates (x^i), the value m_λ^2 independently of the coordinates used. We shall interpret this value as the square of the modulus of linear deformation in a representation, which we shall call affine, of Σ onto another surface $\bar{\Sigma}$, as yet indeterminate. Equation (9′) then states that the tensor a_{ij} is the fundamental metric tensor on the new surface, when it is referred to the same coordinates (x^i) as used on Σ, or when the correspondence of points between Σ and $\bar{\Sigma}$ is effected by equality of coordinates.

Let us denote by A^{ij} the reciprocal elements of a_{ij} in the matrix of the tensor a_{ij}

$$a_{ij} A^{jh} = \delta_i^h; \tag{10}$$

the A^{ij} are then in turn the components of a double contravariant tensor. Setting

$$A = | a_{ij} |; \qquad \frac{1}{A} = | A^{ij} |; \tag{11}$$

and further

$$a_j^r = g^{rh} a_{hj}; \qquad a^{rs} = g^{rh} g^{ks} a_{hk} \tag{12}$$

accepting, from now on, that all the operations of raising and lowering of indices, as defined in the tensor calculus, and thus all the operations of covariant differentiation, take place with respect to the metric g_{ij} of Σ.

Thus the operation of covariant differentiation carried out on the tensor a_{ij} will be, for example,

$$a_{ij/r} = \frac{\partial a_{ij}}{\partial x^r} - a_{is} \left\{ \begin{matrix} s \\ j \quad r \end{matrix} \right\} - a_{sj} \left\{ \begin{matrix} s \\ i \quad r \end{matrix} \right\}; \tag{13}$$

the first invariant of a_{ij} will then be given by

$$\mathfrak{A} = g^{ij} a_{ij} = m_1^2 + m_2^2; \tag{14}$$

and is thus equal to the sum of the squares of the principal moduli of deformation.

With the tensor a_{ij} we can construct the Christoffel symbols of the first and second kinds on $\bar{\Sigma}$:

$$(\overline{ij,h}) = (\overline{ji,h}) = \frac{1}{2}\left(\frac{\partial a_{hi}}{\partial x^j} + \frac{\partial a_{jh}}{\partial x^i} - \frac{\partial a_{ij}}{\partial x^h}\right) \tag{15}$$

$$\overline{\begin{Bmatrix} k \\ i \quad j \end{Bmatrix}} = \overline{\begin{Bmatrix} k \\ j \quad i \end{Bmatrix}} = A^{hk}(\overline{ij,h}) \tag{16}$$

and hence

$$\overline{\begin{Bmatrix} j \\ i \quad j \end{Bmatrix}} = \frac{\partial}{\partial x^i} \lg \sqrt{A}\,. \tag{17}$$

We note further that

$$S = e^{\sigma} = \sqrt{\frac{A}{G}} \tag{18}$$

represents the modulus of areal deformation of the representation of Σ on $\bar{\Sigma}$. Moreover, the following relationships hold between the tensors ε_{ij} and $\bar{\varepsilon}_{ij}$ relative to the two surfaces

$$\varepsilon_{ij} = \sqrt{G}\, e_{ij}\,; \qquad \bar{\varepsilon}_{ij} = \sqrt{A}\, e_{ij}\,; \qquad \bar{\varepsilon}_{ij} = S\varepsilon_{ij} = e^{\sigma}\varepsilon_{ij} \tag{19}$$

where $e_{12} = -e_{21} = 1\,;\ e_{11} = e_{22} = 0$.

4. Given this, let us investigate the relationships that connect the Christoffel symbols of the second kind relative to the two surfaces. With this aim, let us substitute in Eq. (15) in place of the partial derivatives of the components of the tensor a_{ij}, their covariant derivatives with respect to the metric of Σ, obtained from Eq. (13). It is readily found, taking account of Eqs. (10) and (16), that

$$\overline{\begin{Bmatrix} h \\ i \quad j \end{Bmatrix}} = \begin{Bmatrix} h \\ i \quad j \end{Bmatrix} + T^h_{ij} \tag{20}$$

where

$$T^h_{ij} = \frac{1}{2} A^{hr}(a_{ri/j} + a_{jr/i} - a_{ij/r}) = T^h_{ji}\,. \tag{21}$$

The Christoffel symbols of the second kind differ then by a tensor, which thus characterises the difference existing between the connexion of the surface Σ and that of $\bar{\Sigma}$.

The tensor character of the difference between the Christoffel symbols of the second kind relative to two different metrics referred to the same analytic manifold, had already been recognised by Levi-Civita in his *Lezioni di Calcolo differenziale assoluto* (Levi-Civita 1925) translated later into English (Levi-Civita 1947); the tensor T^h_{ij} was then considered by Eddington (1952) in his studies on relativity, with the intention of generalizing the connexion studied by H. Weyl, and which bears his name, and has been taken up again by Schouten (1954) in his investigations on more general connexions.

Contracting Eq. (20) with respect to the indices h, j, and taking note of Eqs. (8), (17) and (18) we then have

$$T^j_{ij} = \sigma/_i\,; \qquad \sigma/_i = \frac{\partial \sigma}{\partial x^i}\,; \tag{22}$$

a relationship that links the tensor T^h_{ij} with the gradient of the logarithm of the modulus of areal deformation of the representation.

Let us note further that, compounding the tensor T^h_{ij} with the fundamental metric tensor of Σ in its contravariant form, we obtain a new vector $\boldsymbol{\tau}$ of contravariant components τ^h

$$\tau^h = g^{ij} T^h_{ij} \tag{23}$$

which is an intrinsic element of the representation. Specifically, taking note of Eq. (21), we have

$$\begin{aligned} \tau^h &= g^{ij} T^h_{ij} = A^{hr}\left(a_{ri}/^i - \frac{1}{2} a^i_i/_r\right) \\ &= A^{hr}\left(\operatorname{div}_r a - \frac{1}{2}\mathfrak{A}/_r\right) = g^{ij}\overline{\begin{Bmatrix} h \\ i\ j \end{Bmatrix}} + \frac{1}{\sqrt{G}}\frac{\partial}{\partial x^i}\sqrt{G}\, g^{hi}\,. \end{aligned} \tag{24}$$

The first term in parenthesis represents the divergence of the tensor a_{ij}; the second, the half of the gradient of its first invariant; and thus, as has been seen, half of the gradient of the sum of squares of the principal moduli of deformation of the representation.

5. Having thus determined the Christoffel symbols of the second kind of $\bar{\Sigma}$, we can, with their help, calculate the Riemann-Christoffel tensor for $\bar{\Sigma}$, and set this in relation to the analogous tensor for Σ.

We have, as is well known,

$$\bar{R}_{i\cdot kr}^{\ j} = \frac{\partial}{\partial x^r}\overline{\begin{Bmatrix} j \\ i\ k \end{Bmatrix}} - \frac{\partial}{\partial x^k}\overline{\begin{Bmatrix} j \\ i\ r \end{Bmatrix}} - \overline{\begin{Bmatrix} j \\ h\ r \end{Bmatrix}}\,\overline{\begin{Bmatrix} h \\ i\ k \end{Bmatrix}} - \overline{\begin{Bmatrix} j \\ h\ k \end{Bmatrix}}\,\overline{\begin{Bmatrix} h \\ i\ r \end{Bmatrix}}\,; \tag{25}$$

taking account of Eq. (20), and of

$$\frac{\partial}{\partial x^r} T^j_{ik} = T^j_{ik/r} + T^j_{is}\begin{Bmatrix} s \\ k\ r \end{Bmatrix} + T^j_{sk}\begin{Bmatrix} s \\ i\ r \end{Bmatrix} - T^s_{ik}\begin{Bmatrix} j \\ r\ s \end{Bmatrix} \tag{26}$$

we have the following relationship

$$\bar{R}_{i\cdot kr}^{\ j} = R_{i\cdot kr}^{\ j} + 2\,T^j_{i[k/r]} - 2\,T^j_{h[k} T^h_{r]i}\,, \tag{27}$$

where the square brackets serve to indicate alternation of the indices enclosed by them, e.g.,

$$T^j_{i[k/r]} = \frac{1}{2}(T^j_{ik/r} - T^j_{ir/k})\,. \tag{28}$$

Equation (27) has already been given by Levi-Civita (1925, p. 224) and by Schouten (1954, p. 138 et seq.).

Contracting Eq. (27) with respect to the indices j, k we obtain the contracted tensor of Riemann, symmetric in the two indices i, r

$$\bar{R}_{ir} = \bar{R}\,{}_{i\cdot jr}^{\ j} = R_{ir} + \sigma/_{ir} - \operatorname{div}_{ir} T - \sigma/_h T_{ir}^h + T_{hr}^j T_{ij}^h; \tag{29}$$

where

$$\sigma/_{ir} = T_{ij/r}^j \tag{30}$$

represents the second covariant derivative of σ with respect to x^i and x^r (in that order), and

$$\operatorname{div}_{ir} T = T_{ir/j}^j \tag{31}$$

represents the divergence of the tensor T_{ir}^j (taken with respect to the contravariant index).

Compounding Eq. (29) with the fundamental metric tensor of Σ in the contravariant form, we have then

$$g^{ir}\bar{R}_{ir} = -2\mathfrak{K} + \Delta_2\sigma - \operatorname{div}\boldsymbol{\tau} - \boldsymbol{\tau}\cdot\operatorname{grad}\sigma + g^{ir}T_{hr}^j T_{ij}^h \tag{32}$$

where

$$g^{ir}\sigma/_{ir} \quad = \operatorname{div}\operatorname{grad}\sigma = \Delta_2\sigma \tag{33}$$

$$g^{ir}\operatorname{div}_{ir} T = g^{ir}T_{ir/j}^j \quad = \tau^j/_j = \operatorname{div}\boldsymbol{\tau} \tag{33'}$$

$$g^{ir}\sigma/_h T_{ir}^h = \sigma/_h \tau^h \quad = \boldsymbol{\tau}\cdot\operatorname{grad}\sigma \tag{33''}$$

and finally

$$R_{ir} = -g_{ir}\mathfrak{K} \tag{33'''}$$

where $\mathfrak{K}$ is the total curvature of Σ; hence

$$g^{ir}R_{ir} = -2\mathfrak{K}\,. \tag{33''''}$$

We recall now that in the same way

$$\bar{R}_{ir} = -a_{ir}\bar{\mathfrak{K}} \tag{34}$$

and that

$$g^{ir}a_{ir} = \mathfrak{A} \tag{35}$$

so that we arrive at the last formula

$$\mathfrak{A}\bar{\mathfrak{K}} = 2\mathfrak{K} - \Delta_2\sigma + \operatorname{div}\boldsymbol{\tau} + \boldsymbol{\tau}\cdot\operatorname{grad}\sigma - g^{ir}T_{hr}^j T_{ij}^h \tag{36}$$

which relates the total curvatures of two surfaces in an affine representation.

6. Another problem of great importance in the theory of cartographic representations is that of determining the tangential curvature $\bar{\gamma}$ of the transform $(\bar{c})$ on $\bar{\Sigma}$ of a curve (c) of tangential curvature γ given on Σ.

Denoting by δ the symbol of absolute differentiation

$$\delta\lambda^i = \frac{\delta\lambda^i}{\delta s}\,ds = \lambda^i/_j\,dx^j = \lambda^i/_j\,\lambda^j ds\,, \tag{37}$$

we have, as is known

$$\text{on } \Sigma: \frac{\delta \lambda^i}{\delta s} = \gamma \nu^i; \qquad \text{on } \bar{\Sigma}: \frac{\delta \bar{\lambda}^i}{\delta \bar{s}} = \bar{\gamma} \bar{\nu}^i \tag{38}$$

where ν^i, $\bar{\nu}^i$ are the contravariant components of the versor of the normal (with appropriate orientation in relation to the sign of γ) to the given curve (c), and to its image $(\bar{c})$, respectively. It is obvious that s represents the arc of the curve (c) considered, as $\bar{s}$ represents that of the transform $(\bar{c})$, and we have $\frac{d\bar{s}}{ds} = m_\lambda$.

Since $\frac{d\lambda^i}{ds} = \frac{d^2x^i}{ds^2}$; $\frac{d\bar{\lambda}^i}{d\bar{s}} = \frac{d^2x^i}{d\bar{s}^2} = \frac{1}{m_\lambda^2} \frac{d^2x^i}{ds^2}$

(m_λ^2 is the square of the modulus of deformation in the direction of the versor $\boldsymbol{t}$ of components λ^i), we have

$$\begin{aligned} \frac{\delta \lambda^i}{\delta s} &= \frac{d^2x^i}{ds^2} + \lambda^r \lambda^s \left\{ \begin{matrix} i \\ r \;\; s \end{matrix} \right\} \\ \frac{\delta \bar{\lambda}^i}{\delta \bar{s}} &= \frac{d^2x^i}{d\bar{s}^2} + \bar{\lambda}^r \bar{\lambda}^s \left\{ \overline{\begin{matrix} i \\ r \;\; s \end{matrix}} \right\} = \frac{1}{m_\lambda^2} \left(\frac{d^2x^i}{ds^2} + \lambda^r \lambda^s \left\{ \overline{\begin{matrix} i \\ r \;\; s \end{matrix}} \right\} \right) \end{aligned} \tag{39}$$

whence

$$m_\lambda^2 \bar{\gamma} \bar{\nu}^i - \gamma \nu^i = T_{rs}^i \lambda^r \lambda^s . \tag{40}$$

Compounding with $\bar{\nu}^i$ we then have

$$\bar{\gamma} = \frac{1}{m_\lambda^2} (\gamma \nu^i \bar{\nu}_i + T_{rs}^i \lambda^r \lambda^s \bar{\nu}_i) . \tag{41}$$

Now

$$\bar{\nu}_i = \bar{\varepsilon}_{ji} \bar{\lambda}^j = \frac{S}{m_\lambda} \varepsilon_{ji} \lambda^j \tag{42}$$

and hence

$$\bar{\gamma} = \frac{S}{m_\lambda^3} (\gamma - \varepsilon_{ji} T_{rs}^i \lambda^j \lambda^r \lambda^s) \tag{43}$$

a formula which I have already published (Marussi 1954).

Only the symmetric part of the new tensor $\varepsilon_{ji} T_{rs}^i$ is involved in this formula, and we denote it by γ_{jrs}

$$\gamma_{jrs} = -\varepsilon_{i(j} T_{rs)}^i . \tag{44}$$

Its components are obtained directly as follows

$$\begin{aligned} \gamma_{111} &= \sqrt{G}\, T_{11}^2 ; \qquad \gamma_{112} = \gamma_{121} = \gamma_{211} = -\frac{1}{3} \sqrt{G} (T_{11}^1 - 2 T_{12}^2) \\ \gamma_{222} &= -\sqrt{G}\, T_{22}^1 ; \qquad \gamma_{122} = \gamma_{212} = \gamma_{221} = \frac{1}{3} \sqrt{G} (T_{22}^2 - 2 T_{12}^1) . \end{aligned} \tag{45}$$

7. Let us now apply the above results to the cases of conformal, equivalent, and geodetic representations.

In the case of *conformal representations,* the tensor a_{ij} is isotropic, in as much as the modulus of deformation is independent of the direction; we have then

$$a_{ij} = e^{2\mu} g_{ij} \tag{46}$$

where μ is the logarithm of the modulus of linear deformation. We have then, as is well known

$$T^{i}_{rs} = \delta^{i}_{r}\mu_{/s} + \delta^{i}_{s}\mu_{/r} - g_{rs}\mu^{/i} \tag{47}$$

whence

$$T^{i}_{ri} = \sigma_{/r} = 2\mu_{/r}\,; \qquad \tau^{i} = 0\,. \tag{48}$$

Formula (27), which relates the Riemann-Christoffel tensor of the two kinds in a conformal representation, was given in 1903 by Finzi.

The relationship which links the total curvatures of the two surfaces is then written

$$e^{2\mu}\bar{\mathfrak{K}} = \mathfrak{K} - \Delta_2\mu \tag{49}$$

are already discussed in my notes (Marussi 1951a, b). It was given for the first time, as far as I know, by Souslow (1898).

The relationship which links the tangential curvatures of two curves in a conformal representation is finally written as

$$\bar{\gamma} = e^{-\mu}(\gamma - \mu^{/i} v_i)\,, \tag{50}$$

and hence expresses the well known theorem of Schols, concerning which see the note of Taucer (1954). In this last equation the term $\mu^{/i} v_i$ obviously expresses the specific flux of the gradient of the logarithm of the modulus of deformation, across the line (c) at the point considered.

8. In equivalent representations, σ is independent of the coordinates. We then have, from Eq. (22)

$$\sigma_{/r} = T^{h}_{hr} = \frac{1}{2} A^{ih} a_{ih/r} = 0 \tag{51}$$

whence Eq. (36) is written

$$\mathfrak{A}\,\bar{\mathfrak{K}} = 2\,\mathfrak{K} + \operatorname{div}\tau - g^{ir} T^{j}_{hr} T^{h}_{ij}\,; \tag{52}$$

for which the tensor γ_{ijh} takes on a particularly simple form

$$\begin{aligned} \gamma_{111} &= \sqrt{G}\,T^{2}_{11}\,; \qquad \gamma_{112} = \gamma_{121} = \gamma_{211} = -\sqrt{G}\,T^{1}_{11} \\ \gamma_{222} &= -\sqrt{G}\,T^{1}_{22}\,; \qquad \gamma_{122} = \gamma_{212} = \gamma_{221} = \sqrt{G}\,T^{2}_{22}\,. \end{aligned} \tag{53}$$

9. Finally, in *geodetic representations,* the tensor γ_{ijh} is zero, and this leads to the following relationships, taking account of (45),

$$T^{2}_{11} = T^{1}_{22} = 0\,; \qquad T^{1}_{11} = 2\,T^{2}_{12}\,; \qquad T^{2}_{22} = 2\,T^{1}_{12}\,. \tag{54}$$

The relationships between the Christoffel symbols relative to two metrics in an affine transformation have been given by Dini (1869) and are cited in the treatise of Bianchi (1927) in relation to the general problem of geodetic representations.

Since in this case only two scalars suffice to completely define the tensor T^i_{rs}, we can express the latter by means of the components of a vector p_i; and we see that the Eqs. (54) are satisfied if we put (Eisenhart 1926)

$$T^i_{rs} = \delta^i_s p_r + \delta^i_r p_s \,. \tag{55}$$

The geometric significance of the vector p_i is immediately seen by contracting (55)

$$T^i_{ri} = \sigma_{/r} = 3\,p_r \tag{56}$$

whence we shall write

$$T^i_{rs} = \frac{1}{3}(\delta^i_s \sigma_{/r} + \delta^i_r \sigma_{/s}) \,. \tag{57}$$

For the components of the vector τ^i we would then have

$$\tau^i = \frac{2}{3}\sigma^{/i} \tag{58}$$

whence

$$\tau = \frac{2}{3}\operatorname{grad}\sigma \,. \tag{59}$$

The relationship between the total curvatures of the two surfaces in a geodetic representation then takes on a particularly simple form

$$\mathfrak{A}\bar{\mathfrak{K}} = 2\,\mathfrak{K} - \Delta_2 p + \operatorname{grad}^2 p \,; \qquad p = \frac{1}{3}\sigma \,. \tag{60}$$

Bibliography

Bianchi L (1927) Lezioni di geometria differenziale, vol I, part 1. Zanichelli, Bologna, pp 313

Dini U (1869) Sopra un problema che si presenta nella teoria generale della rappresentazione geografica di una superficie sopra un'altra. Ann Mat Ser 1, III

Eddington AS (1952) The mathematical theory of relativity, Chap. VII. Cambridge Univ Press, Cambridge

Eisenhart LP (1926) Riemannian Geometry. Princeton Univ Press, Princeton, p 132

Finzi A (1903) Le ipersuperficie a tre dimensioni che si possono rappresentare conformemente nello spazio euclideo. Atti R Ist Veneto 62:1049 – 1062 (also Levi-Civita 1947, p 231)

Levi-Civita T (1925) Lezioni di calcolo differenziale assoluto. Stock, Rome

Levi-Civita T (1947) The absolute differential calculus (calculus of tensors). Blackie, London Glasgow, pp 220

Marussi A (1951a) Su alcune proprietà fondamentali delle rappresentazioni conformi fra superfici qualunque. Boll Geodes Sci Affini no 2

Marussi A (1951b) Su alcune proprietà integrali delle rappresentazioni conformi di superfici su superfici. Rend Accad Naz Lincei 10: Ser 8, fasc 4 (Chap. V.1, this Vol.)

Marussi A (1954) Sulla curvatura tangenziale delle trasformate di curve nelle rappresentazioni affini fra superfici. Rend Accad Naz Lincei 16: Ser 8 fasc 4
Schouten JA (1954) Ricci calculus. Springer, Berlin Göttingen Heidelberg
Souslow G (1898) Sur la représentation conforme d'une surface sur une autre. C R Acad Sci 126
Taucer G (1954) Alcune considerazioni sul teorema di Schols. Boll Geodes Sci Affini no 2

VI Propagation of Light in Continuous Isotropic Refracting Media

1 An Analogy Between the Laws of Propagation of Light in Continuous Isotropic Refracting Media and Conformal Representations[1]

1. The simple observation that, denoting by

$$m = \frac{ds}{d\bar{s}} \tag{1}$$

the scale factor in a conformal representation between three-dimensional spaces S and $\bar{s}$, and considering a point $\bar{P}$ moving in $\bar{S}$ with velocity $c = \dfrac{d\bar{s}}{dt}$, the velocity v of its image P in $\bar{S}$ will be

$$v = \frac{ds}{dt} = mc \tag{2}$$

demonstrates the strict analogy between the properties of these representations, and the laws of propagation of light in an isotropic refracting medium, by identifying the scale factor m with the reciprocal of the index of refraction n of such a medium. In the case of light there is in fact a quite analogous formula.

$$v = \frac{c}{n} \tag{2'}$$

where c is the velocity of propagation *in vacuo*; and both m and n are point functions, not depending on direction.

This observation can find a more precise formulation if we recall Fermat's variational principle, according to which a stationary optical path corresponds to the trajectory effectively followed by the light, and hence

$$\delta \int_A^B dt = \frac{1}{c} \delta \int_A^B n\, ds = 0 \tag{3}$$

in which A and B are two points on the path. We also note that an analogous principle holds for the images of geodesics in the conformal representation under consideration. In fact, for the geodesic $\bar{g}$ of $\bar{S}$ passing through two points $\bar{A}$ and $\bar{B}$ we have

$$\delta \int_{\bar{A}}^{\bar{B}} d\bar{s} = 0 \tag{4}$$

[1] Originally published as: Marussi A (1953) Un'analogia fra le leggi della propagazione della luce in mezzi rifrangenti continui isotropi, e le rappresentazioni conformi. Atti Fond Giorgio Ronchi Ist Naz Ottica, anno VIII: no 1

in which the line integral is taken along $\bar{g}$; and hence, by Eq. (1) we still have likewise

$$\delta \int_A^B \frac{ds}{m} = 0 \tag{4'}$$

along g, A and B being the points corresponding to $\bar{A}$ and $\bar{B}$ in the representation.

There is thus a perfect analogy between Eqs. (3) and (4'); and hence it will always be possible to consider the light paths in a continuous isotropic refracting medium of index n, as the images of geodesics in a conformal representation of scale factor $m = \frac{1}{n}$, and vice versa.

2. Following this, several properties of conformal representations have immediate equivalents in optics. To see this, let us look at the differential equation which is satisfied by the images of geodesics in a conformal representation, which we shall take to be characterized by its scale factor m understood to be a point function, or better by its natural logarithm:

$$\mu = \lg m\,, \tag{5}$$

whence this equation will be obtained by substituting in Eq. (4') for Euler's formula. To this end we shall call $\bar{a}_{ij}$ and $a_{ij} = e^{2\mu}\bar{a}_{ij}\,(i,j = 1,2,3)$ respectively the fundamental metric tensors of $\bar{S}$ and S referred to the same coordinates x^i provided that the same coordinates apply to the corresponding points P and $\bar{P}$. The (contravariant) components of the versor tangent to g will then be denoted by $\lambda^i = \frac{\partial x^i}{\partial s}$; and Eq. (4') can now be written so as to make clear that the curve along which we integrate is the one considered

$$\delta \int_A^B e^{-\mu} \sqrt{a_{ij}\lambda^i\lambda^j}\, ds = \delta \int_A^B \Phi ds\,; \qquad \Phi = e^{-\mu}\sqrt{a_{ij}\lambda^i\lambda^j} \tag{6}$$

since $ds^2 = a_{ij}\lambda^i\lambda^j ds^2$. (Summation over repeated indices).

Euler's formulae then give the system

$$\frac{d}{ds}\left(\frac{\partial \Phi}{\partial \lambda^i}\right) - \frac{\partial \Phi}{\partial x^i} = 0\,. \qquad (i = 1,2,3) \tag{7}$$

Expanding the calculations we have, putting $\mu_i = \frac{\partial \mu}{\partial x^i}$:

$$\frac{\partial \Phi}{\partial \lambda^i} = e^{-\mu} a_{ij}\lambda^j$$

$$\frac{\partial \Phi}{\partial x^i} = e^{-\mu}\left(\frac{1}{2}\,\frac{\partial a_{jh}}{\partial x^i}\,\lambda^j\lambda^h - \mu_i\right)$$

$$\frac{d}{ds}\left(\frac{\partial \Phi}{\partial \lambda^i}\right) = e^{-\mu}\left(a_{ij}\frac{d\lambda^j}{ds} + \frac{\partial a_{ij}}{\partial x^k}\,\lambda^j\lambda^k - \mu_h a_{ij}\lambda^j\lambda^h\right) \tag{8}$$

and Eq. (7) becomes

$$a_{ij}\frac{d\lambda^j}{ds}+(jk,i)\lambda^j\lambda^k=-(\mu_i-\mu_k\lambda^k\lambda_i) \tag{9}$$

where (jk, i) is Christoffel's symbol of the first kind calculated for the metric a_{ij}, and $\lambda_i=a_{ij}\lambda^j$.

We now note that the expressions

$$\mu_i-\mu_k\lambda^k\lambda_i=(\mu_i)_n \tag{10}$$

represent the covariant tensor components of the vector component normal to the curve g of the gradient of μ (which we shall call gradient of deformation); but from (9), on the other hand, it follows that this component equals the vector of curvature of g; whence: the osculating plane of the image of a geodesic contains the gradient of deformation; its first curvature equals the modulus of its normal component; this latter is then found on the convex side of the curve.

In the case of spaces of two dimensions (surfaces), this formulation is modified in the well known theorem of Schols: the tangential curvature of the image of a geodesic in a conformal representation between surfaces equals the modulus of the normal component of the gradient of deformation (or its unitary flux across the curve); the gradient of deformation lies on the convex side of the curve.

3. In the optical analogy, putting ν as the logarithm of n

$$\nu=\lg n \tag{11}$$

we have

$$\nu=-\mu;\qquad \operatorname{grad}\nu=-\operatorname{grad}\mu; \tag{12}$$

the properties corresponding to those which have just been set out for conformal representations are well known: the osculating plane to a light path contains the normal to the surfaces of equal index of refraction; the first curvature of the path equals the modulus of the normal component of the gradient of the logarithm of the index of refraction, and this lies on the concave side of the path.

We have thus established the reciprocity between optical phenomena and conformal representations; but, given a refracting medium of index n, in general there is no conformal correspondence between Euclidean spaces which can have $1/n$ as scale factor. For this to hold, the image space will in general be Riemannian.

Thus if we consider a plane refracting medium, indicating as usual by $\nu=\mu$ the logarithm of the index of refraction, it so happens that the surface which should be represented conformally on the plane with the given scale factor, would have a total curvature K related to μ or to ν by the relationship

$$K=-e^{-2\mu}\Delta_2\mu=e^{-2\nu}\Delta_2\nu \tag{13}$$

the Δ_2 (Laplacian) being calculated on the given plane. More complex relationships apply for the appropriate Riemann-Christoffel tensor in the case of three-dimensional spaces (Eisenhart 1949, Levi-Civita 1947, Finzi 1902–3, Marussi 1951).

4. Another analogy can be established between the phenomena of refraction, conformal representations, and mechanics, by recalling the variational principle of Hamilton-Hölder, expressed by the equation:

$$\delta \int_A^B \sqrt{2(U+E)}\, ds = 0\,, \tag{14}$$

the line integral being extended along the natural trajectory of a moving body subject to a force field with a potential U, and given that $E = T - U$ is the energy constant specific to the motion considered (T is the kinetic energy).

If we consider the bundle of trajectories for which the energy constant is zero, $E = 0$, then Eq. (14) is written

$$\delta \int_A^B \sqrt{2U}\, ds = 0 \tag{14'}$$

and such a bundle will coincide with that of the images of the geodesics, or if we wish, with that of the optical paths from the duality established above, when (Levi-Civita and Amaldi 1927)

$$\sqrt{2U} = \frac{1}{m} = n\,; \qquad U = \frac{1}{2m^2} = \frac{n^2}{2}\,. \tag{15}$$

And hence: In a conformal transformation of scale factor m the images of the geodesics are the same as the dynamic trajectories of a material point subjected to a force derived from a potential $U = 1/2m^2$, and characterised by a zero value of the energy constant E; or as the paths of a light ray propagated in an isotropic medium of refractive index $n = 1/m$.

Bibliography

Eisenhart LP (1949) Riemannian geometry. Princeton Univ Press, Princeton, p 89

Finzi A (1902 – 3) Le ipersuperfici a tre dimensioni che si possono rappresentare conformemente nello spazio euclideo. Atti R Ist Veneto Sci Lett Art, LXII, part II, p 1049

Levi-Civita T (1947) The absolute differential calculus. Blackie, London Glasgow, p 228

Levi-Civita T, Amaldi U (1927) Lezioni di meccanica razionale, vol II, part II. Zanichelli, Bologna, p 509, 515

Marussi A (1951) Su alcune proprietà fondamentali delle rappresentazioni conformi fra superfici. Boll Geodes Sci Affini no 2

2 The Torsion of a Ray of Light in a Continuous Isotropic Refracting Medium [1]

1. The vertical plane at a point A of the Earth's surface, also containing another point B on it, is given form in Geodesy by means of the light path which joins such points, and since the path is in general curved, the exact definition of such a vertical plane depends:

a) in the case when the light path can be considered plane, on the inclination of its plane with respect to the vertical plane considered;
b) in the case when the light path cannot be considered as plane, on its torsion, as well as on the inclination of its osculating plane at A, with respect to the vertical plane considered.

The necessity of defining rigorously the geometric elements which are involved in this geodetic problem forms the point of departure for the research which is summarised in the present chapter; and, as far as I know, these have never been treated in an organic and general manner, despite their simplicity.

These geometric elements obviously depend on the distribution of the refractive index in space; their determination is obtained by an extension of the known relationships between the gradient of the refractive index, and the curvature vector of the light path, which are consequences of Fermat's principle (Chap. VI.1, this Vol.).

2. Let us suppose that we are given the refractive index n, or, better, its logarithm

$$v = \lg n \tag{1}$$

as a continuous point function; and we shall further consider its gradient

$$\operatorname{grad} v = fN \tag{2}$$

in which f is thus its modulus (essentially positive), and N its versor. N therefore represents also the normal to the surface Σ of constant refractive index (the equirefrangent surface), directed in the sense of increasing refractive index.

Since we shall need to consider the derivatives in various directions of the quantities which we have defined until now, we shall make use of the very concise method of representing these by means of vectorial homographies. Taking a general point P, we denote by σ the homography

[1] Originally published as: Marussi A (1953) La torsione di una traiettoria luminosa in un mezzo rifrangente continuo isotropo. Atti Fond Giorgio Ronchi Ist Naz Ottica, anno VIII: no 2, March to April 1953

$$\sigma = \frac{dN}{dP} \tag{3}$$

and by σ_Σ the surface homography on Σ, which for this surface is identified with the well-known homography of Burali-Forti.

We thus have

$$\sigma = \sigma_\Sigma + H(N, \sigma N) \tag{4}$$

where $H(\boldsymbol{a}, \boldsymbol{b})$ denotes the diad of Gibbs, which, when applied to an arbitrary vector $\boldsymbol{x}$, yields $\boldsymbol{a} \times \boldsymbol{x} \cdot \boldsymbol{b}$.

Next, differentiating Eq. (2) we have

$$\frac{d \operatorname{grad} v}{dP} = f\sigma + H(\operatorname{grad} f, N) \tag{5}$$

from which, by taking the first invariant, and denoting by H the mean curvature of Σ, we have

$$\Delta_2 v = fH + \operatorname{grad} f \times N\,. \tag{6}$$

If the logarithm of the refractive index is harmonic ($\Delta_2 v = 0$), it follows that the mean curvature of the equirefrangent surfaces is given by

$$H = -\operatorname{grad} \lg f \times N \; (\Delta_2 v = 0) \tag{7}$$

or otherwise by

$$H = -\operatorname{grad} \lg f \times N + \frac{\Delta_2 v}{f}\,. \tag{7'}$$

3. From the relationship

$$f^2 = (\operatorname{grad} v)^2 \tag{8}$$

differentiating with respect to the point P, and recalling that the homography $\frac{d \operatorname{grad} v}{dP}$ is a dilatation, we have

$$f \operatorname{grad} f = \frac{d \operatorname{grad} v}{dP} \operatorname{grad} v \tag{9}$$

whence, from Eq. (2)

$$\frac{d \operatorname{grad} v}{dP} N = \operatorname{grad} f \tag{10}$$

and therefore, taking account of Eq. (5)

$$\sigma N = \operatorname{grad}_\Sigma \lg f \tag{11}$$

having denoted by $\operatorname{grad}_\Sigma \lg f$ the component of $\operatorname{grad} \lg f$ on the equirefrangent surface Σ, i.e. $\operatorname{grad}_\Sigma \lg f = \operatorname{grad} \lg f - \operatorname{grad} \lg f \times N \cdot N$. Now, we note that σN gives us the curvature vector of the trajectories orthogonal to the equirefrangent surfaces, i.e., the vector directed along the principal normal $\boldsymbol{M}$ to these lines, and having as modulus their curvature of flexion Γ, whence

$$\mathrm{grad}_\Sigma \lg f = \Gamma \boldsymbol{M} . \tag{12}$$

4. Given all this, let us consider a light path (g), and denoting by s its arc, and by $\boldsymbol{t}$ the versor of its tangent:

$$\boldsymbol{t} = \frac{dP}{ds} \tag{13}$$

with $\boldsymbol{n}$ the versor of its principal normal, and $\boldsymbol{b}$ that of its binormal, then $\boldsymbol{b} = \boldsymbol{t} \wedge \boldsymbol{n}$, and the triplet $(\boldsymbol{t}, \boldsymbol{n}, \boldsymbol{b})$ is thus positive (right-handed). We further denote by γ the first curvature or flexion of (g), and by τ its second curvature or torsion, which are governed by the following formulae of Frenet:

$$\frac{d\boldsymbol{t}}{ds} = \gamma \boldsymbol{n} ; \qquad \frac{d\boldsymbol{b}}{ds} = \tau \boldsymbol{n} . \tag{14}$$

In these, γ is essentially positive; τ is positive when $\boldsymbol{b}$ turns in the normal plane in the anticlockwise sense as seen by an observer aligned along $\boldsymbol{t}$.

From Fermat's principle it follows that the curvature vector $\gamma \boldsymbol{n}$ of (g) is equal to the component of the gradient of v normal to $\boldsymbol{t}$ (Chap. VI.1, this Vol.):

$$\gamma \boldsymbol{n} = \mathrm{grad}\, v - \mathrm{grad}\, v \times \boldsymbol{t} \cdot \boldsymbol{t} \tag{15}$$

whence

$$\boldsymbol{n} = \rho(\mathrm{grad}\, v - \mathrm{grad}\, v \times \boldsymbol{t} \cdot \boldsymbol{t}) \tag{16}$$

where we have denoted by $\rho = \dfrac{1}{\gamma}$ the radius of first curvature of the light paths.

We thus have

$$\boldsymbol{b} = \boldsymbol{t} \wedge \boldsymbol{n} = \rho \boldsymbol{t} \wedge \mathrm{grad}\, v = \rho f \boldsymbol{t} \wedge \boldsymbol{N} . \tag{17}$$

Differentiating this with respect to s, we then have

$$\frac{d\boldsymbol{b}}{ds} = \frac{d}{ds}(\rho f) \boldsymbol{t} \wedge \boldsymbol{N} + f \boldsymbol{n} \wedge \boldsymbol{N} + \rho f \boldsymbol{t} \wedge \sigma \boldsymbol{t} \tag{18}$$

since

$$\frac{d\boldsymbol{N}}{ds} = \frac{d\boldsymbol{N}}{dP} \frac{dP}{ds} = \sigma \boldsymbol{t} \tag{19}$$

and therefore

$$\tau = \frac{d\boldsymbol{b}}{ds} \times \boldsymbol{n} = \rho f \boldsymbol{n} \times \boldsymbol{t} \wedge \sigma \boldsymbol{t} = \rho f \boldsymbol{n} \wedge \boldsymbol{t} \times \sigma \boldsymbol{t} = -\rho f \sigma \boldsymbol{t} \times \boldsymbol{b} . \tag{20}$$

This latter gives us the torsion of the light path that we are seeking, but it can also be put in a more explicit form by recalling Eqs. (4) and (11):

$$\sigma \boldsymbol{t} \times \boldsymbol{b} = \sigma_\Sigma \boldsymbol{t} \times \boldsymbol{b} + \boldsymbol{N} \times \boldsymbol{t} \cdot \mathrm{grad}_\Sigma \lg f \times \boldsymbol{b} . \tag{21}$$

If we now denote by θ the angle between the principal normal to the path, and the normal N to the equirefrangent surface (the angle of elevation of the path with respect to the equirefrangent surfaces):

$$\cos\theta = \boldsymbol{N}\times\boldsymbol{n}\,; \qquad \boldsymbol{N}\times\boldsymbol{t} = -\sin\theta \tag{22}$$

from Eq. (15), however, we have

$$\gamma = f\cos\theta; \qquad \rho f = \frac{1}{\cos\theta}. \tag{23}$$

Calling $\boldsymbol{t}'\cos\theta$ the component of $\boldsymbol{t}$ on Σ ($\boldsymbol{t}'$ is thus a unit vector), and observing that

$$\sigma_\Sigma \boldsymbol{t}'\times\boldsymbol{b} = -(\tau_g)\,t \tag{24}$$

we obtain the geodetic torsion τ_g of the surface in the direction considered, and with change of sign we have:

$$\tau = \tau_g + \frac{\sin\theta}{\gamma}\,\mathrm{grad}_\Sigma f\times\boldsymbol{b} = \tau_g + \Gamma\,\mathrm{tg}\,\theta\sin\psi \tag{25}$$

in which ψ is the angle, measured in a clockwise sense for an observer aligned along $\boldsymbol{N}$, contained between the osculating plane to the orthogonal trajectories of the equirefrangent surfaces, and that to the light path.

As can thus be seen, the torsion of the light path is composed of the sum of two parts: one which depends on the geodetic torsion of the equirefrangent surfaces, in the direction determined on these by the normal plane that contains the path itself; the other depends on the curvature of the trajectories orthogonal to the equirefrangent surfaces or, if we prefer, on the distribution on these of the modulus of the gradient of the index; in yet other words, on the defect of parallelism of the equirefrangent surfaces themselves.

VII Posthumous Work

The Motion of a Free Particle and of a Spherical Pendulum in the Microgravitational Field of a Gravitationally Stabilized Satellite in Circular Orbit in a Central Field

ANTONIO MARUSSI and CLAUDIO CHIARUTTINI

Institute of Geodesy and Geophysics, University of Trieste, Italy

Introduction

Marussi (1979, 1982, 1984) has studied the structure and the intrinsic reference system of the residual microgravitational – or tidal – field sensed in a satellite in inertial motion in the gravitational field of a planet. In the present paper we study the motion of a free particle and of a spherical pendulum in this field.

The expression of the tidal field in the canonical reference triad $P_0(x_1, x_2, x_3)$ is:

$$2T = -g(k_1 x_1^2 + k_2 x_2^2 - H x_3^2 + 2 f_1 x_1 x_3 + 2 f_2 x_2 x_3) .$$

P_0 is the centre of mass of the satellite, x_1 and x_2 are oriented along the principal directions of the equipotential surface of the planet field, x_3 is oriented along the line of force, g is the gravity acceleration, k_1 and k_2 are the principal curvatures of the equipotential surface, $H = k_1 + k_2$ is the mean curvature of the same surface, f_1 and f_2 are the x_1 and x_2 components of the curvature vector of the line of force. All the above quantities are referred to the centre of mass P_0.

If the satellite rotates, the centrifugal potential $\Omega^2 R^2/2$ has to be added to T. Ω is the rotation vector, R is the distance from the rotation axis. The resulting potential is:

$$\begin{aligned} 2W = 2T + \Omega^2 R^2 &= (\Omega_2^2 + \Omega_3^2 - g k_1) x_1^2 + (\Omega_1^2 + \Omega_3^2 - g k_2) x_2^2 \\ &\quad + (\Omega_1^2 + \Omega_2^2 + g H) x_3^2 - 2[\Omega_1 \Omega_2 x_1 x_2 + (\Omega_1 \Omega_3 + g f_1) x_1 x_3 \\ &\quad + (\Omega_2 \Omega_3 + g f_2) x_2 x_3] . \end{aligned} \tag{1}$$

We shall assume that the satellite moves in a central field in circular orbit and that its rotation is gravitationally stabilized. Figure 1 shows the geometry of the problem.

In a central field $k_1 = k_2 = k$, $H = 2k$, $f_1 = f_2 = 0$. If the orbit is circular then the square of the angular speed ω_0 of the satellite is equal to the gravitational acceleration times k which is also the curvature of the orbit: $\omega_0^2 = gk$. The gravitationally stabilized rotation implies that $\Omega_1 = \omega_0$, $\Omega_2 = \Omega_3 = 0$. The potential (1) then reduces to

$$2W = 2T + \omega_0^2 R^2 = \omega_0^2 (3z^2 - x^2) . \tag{2}$$

The equipotential surfaces of the field W are two conjugate families of coaxial cylindrical hyperboloids parallel to y axis (Fig. 2).

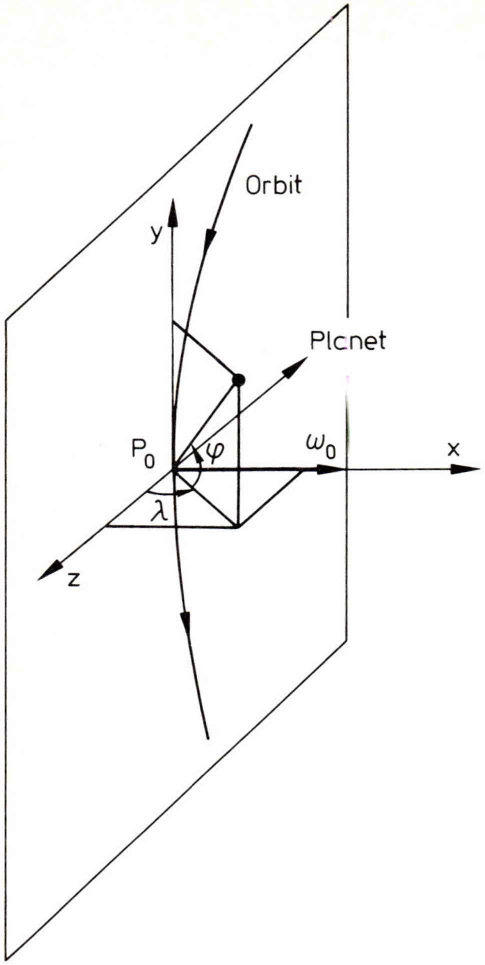

Fig. 1. The geometry of the problem: (y, z) is the orbital plane, the satellite moves in the negative y direction, the axis z is the outward direction in the gravitational field and ω_0 is the gravitationally stabilized rotation vector, φ and λ are the spherical coordinates used in solving the motion of the pendulum

The Free Particle

In a rotating satellite the motion of a body is driven by the conservative force resulting from Eq. (2) and by the force of Coriolis. The Lagrangian for a free particle is:

$$2L = \dot{x}^2 + \dot{y}^2 + \dot{z}^2 + \omega_0^2(3z^2 - x^2) + 2\,\omega_0(y\dot{z} - \dot{y}z)\,. \tag{3}$$

The last term in the right hand side is the one which generates the Coriolis force.

From Eq. (3) the Lagrangian equations of motion are:

$$\ddot{x} + \omega_0^2 x = 0$$

$$\ddot{y} - 2\,\omega_0\dot{z} = 0$$

$$\ddot{z} + 2\,\omega_0\dot{y} - 3\,\omega_0^2 z = 0\,. \tag{4}$$

Fig. 2. Traces on the (x, z) plane of the cylindrical surfaces of the stabilized micro-gravitational field

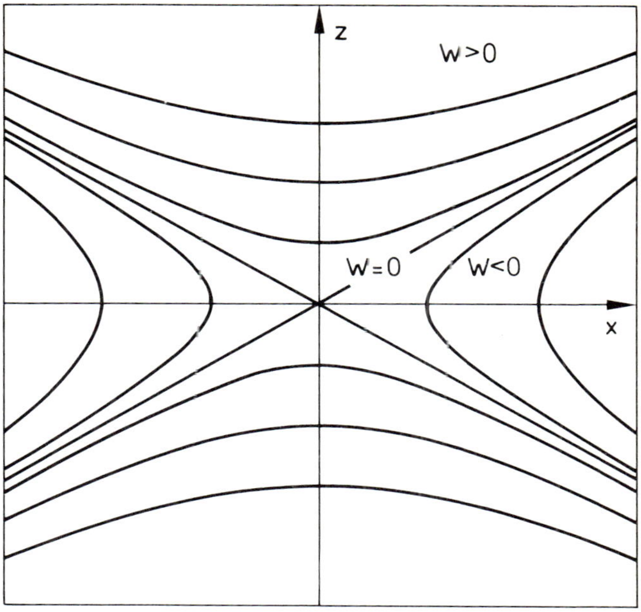

The first equation is independent of the other two and its solution is:

$$x = A_{x1} \exp(i\omega_0 t)$$

with A_{x1} complex.

The last two equations in (4) constitute an independent subsystem in which $\dot{y}$ and z can be considered as the unknown functions. Substituting

$$\dot{y} = V_y \exp(i\omega t)$$

$$z = A_z \exp(i\omega t)$$

into the second and third equation in (4), we obtain the algebraic system

$$\omega V_y - 2\omega\omega_0 A_z = 0$$

$$2\omega_0 V_y - (\omega^2 + 3\omega_0^2) A_z = 0\,, \tag{5}$$

which has non-trivial solutions if its determinant vanishes:

$$-\omega(\omega^2 + 3\omega_0^2) + 4\omega\omega_0^2 = 0\,.$$

The roots are $\omega^2 = 0$ and $\omega^2 = \omega_0^2$.

Substituting $\omega = 0$ into (5) the following relation

$$V_{y0} = 3/2\,\omega_0 A_{z0}$$

is obtained. The index 0 indicates the solution for $\omega = 0$. The solution for the last two equations in (4) in terms of y and z is:

$$y = 3/2\,\omega_0 A_{z0} t + C$$

$$z = A_{z0}$$

i.e.: a constant velocity in the y direction, which depends on the distance A_{z0} from the orbit of the satellite. Here A_{z0} and C are real constants.

Substituting $\omega = \omega_0$ into (5) we obtain the following solution:

$$V_{y1} = 2\,\omega_0 A_{z1} \tag{6}$$

and then

$$\begin{aligned} y &= -2\,iA_{z1}\exp(i\omega_0 t) \\ z &= A_{z1}\exp(i\omega_0 t) \end{aligned} \tag{7}$$

where A_{z1} is complex. The same solution is obtained for $\omega = -\omega_0$. The real parts of (7) furnish the equations of an ellipse with the major axis along y twice the minor axis along z. The particle moves on the ellipse rotating in the direction opposite to the stabilized rotation.

The complete solution of (4) is then

$$\begin{aligned} x &= A_{x1}\exp(i\omega_0 t) \\ y &= -2\,iA_{z1}\exp(i\omega_0 t) + 3/2\,\omega_0 A_{z0} t + C \\ z &= A_{z1}\exp(i\omega_0 t) + A_{z0}\,. \end{aligned}$$

If $A_{z0} \neq 0$ the mass moves on a different orbit than the satellite. As a result, the particle drifts away at a velocity $3/2\,\omega_0 A_{z0}$ relative to the centre of mass of the satellite.

If $A_{z0} = 0$ the trajectory is a closed one, the projections of which on any coordinate plane are ellipses. The y component of the velocity is, from Eq. (6), $2\,\omega_0$ times the distance from the satellite orbit and is thus zero for $z = 0$, as necessary for the mass to be in the same orbit of the satellite.

If the Coriolis force is neglected, dropping the terms in $2\,\omega_0$ in Eq. (4), all three equations are independent of each other. It is worth remarking that in the absence of the Coriolis force the motion would be unbounded in z. The Coriolis force has in fact the effect of stabilizing the system: this is a general property of the gyrostatic forces, i.e. forces which are normal to the velocity and do not contribute to the energy of the system.

The Domain of Motion of the Spherical Pendulum

We assume that the pendulum is bounded to the sphere of radius l of equation

$$x^2 + y^2 + z^2 = l^2 \tag{8}$$

having the centre in the centre of mass of the satellite.

The trajectory of the pendulum is not a simple one; the solution of the equations of motion will be given later for the case of small oscillations. Here we will use energy arguments to determine the domain which contains the trajectory for any amplitude of oscillation.

Since the expression of the kinetic energy is positive definite, the total energy E of system satisfies the inequality $E \geqslant -W$. If we consider E constant and W a function of the coordinates, then

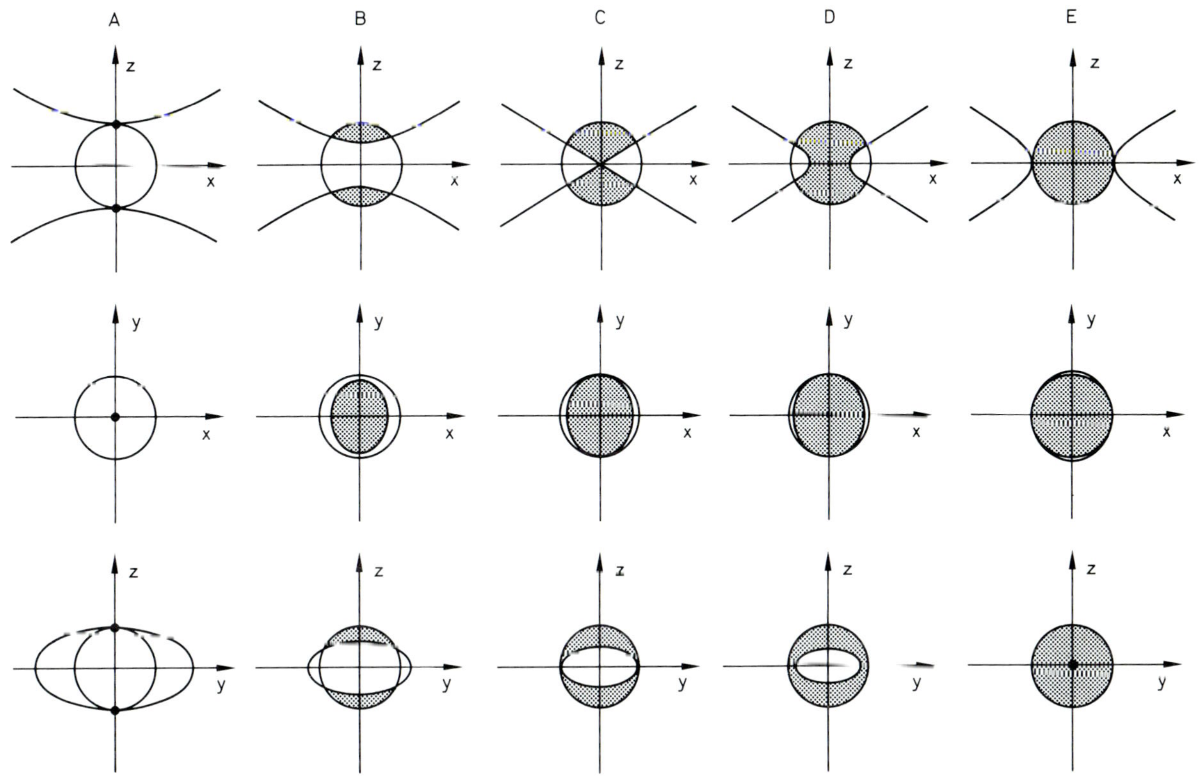

Fig. 3. A – E. Projections on the coordinate planes of the spherical domain containing the trajectory of the pendulum (shaded area), for different values of the energy E. **A** $E = -3/2\,\omega_0^2 l^2$, this is the energy at rest on the z axis $(0, 0, \pm l)$ (stable equilibrium); **B** $E = -3/8\,\omega_0^2 l^2$; **C** $E = 0$, this is the energy at rest on the y axis $(0, \pm l, 0)$ (unstable equilibrium); **D** $E = 1/8\,\omega_0^2 l^2$; **E** $E = 1/2\,\omega_0^2 l^2$, this is the energy at rest on the x axis $(\pm l, 0, 0)$ (unstable equilibrium)

$$W(x, y, z) = \omega_0^2(3z^2 - x^2)/2 \geqslant -E \tag{9}$$

defines a region in the space.

The intersection of the sphere (8) with the region (9) is the domain in which the motion is bounded. Figure 3 shows the projections of the spherical domain so defined on the coordinate planes. The projections on the plane (x, z) are immediately obtained from the equipotential surfaces of the field W. To obtain the projections on the plane (x, y), we eliminate z from the system

$$\begin{aligned} &x^2 - 3z^2 = 2E/\omega_0^2 \\ &x^2 + y^2 + z^2 = l^2 \end{aligned} \tag{10}$$

obtaining the equation

$$4x^2 + 3y^2 = 3l^2 + 2E/\omega_0^2, \tag{11}$$

which represents a family of homothetical ellipses.

The projections of the domain of the motion on the (x, y) plane are the area inside both the ellipses (11) and the circle of radius l.

To obtain the projection on the plane (y, z), eliminating x from the system (10) results in the equation

$$y^2 + 4z^2 = l^2 - 2E/\omega_0^2 \tag{12}$$

which also represents a family of homothetical ellipses.

The projections of the domain of the motion are the area inside the circle of radius l and outside the ellipses (12).

When the pendulum reaches the edge of the allowed domain its velocity is zero.

The Motion of the Pendulum

It is convenient to use a spherical coordinate system in order to simplify the constraint equation. In view of the solution of the equations of motion for small oscillation close to one of the positions of stable equilibrium on the z axis, the y axis will be taken as polar. We call φ the latitude and λ the longitude reckoned positively from the z- towards the x-axis (Fig. 1). The transformations from spherical to Cartesian coordinates are

$$\begin{aligned} x &= r\cos\varphi\sin\lambda \\ y &= r\sin\varphi \\ z &= r\cos\varphi\cos\lambda\,. \end{aligned} \tag{13}$$

By these transformations and by the constraint equation – which is now simply $r = l$ – the Lagrangian becomes:

$$\begin{aligned} 2L = l^2[&\dot{\varphi}^2 + \dot{\lambda}^2\cos^2\varphi + \omega_0^2(3\cos^2\lambda - \sin^2\lambda)\cos^2\varphi \\ &- 2\omega_0(\dot{\varphi}\cos\lambda + \dot{\lambda}\sin\varphi\cos\varphi\sin\lambda)]\,. \end{aligned}$$

The Lagrangian equations of motion are:

$$\ddot{\varphi} + \dot{\lambda}^2 \sin\varphi\cos\varphi + \omega_0^2(3\cos^2\lambda - \sin^2\lambda)\sin\varphi\cos\varphi + 2\,\omega_0\dot{\lambda}\cos^2\varphi\sin\lambda = 0$$

$$\ddot{\lambda}\cos\varphi - 2\,\dot{\varphi}\dot{\lambda}\sin\varphi + 4\,\omega_0^2\cos\varphi\sin\lambda\cos\lambda - 2\,\omega_0\dot{\varphi}\cos\varphi\sin\lambda = 0\,. \tag{14}$$

Since the time does not enter explicitely in the Lagrangian, the system is conservative and the energy integral is:

$$2E = l^2[\dot{\varphi}^2 + \dot{\lambda}^2\cos^2\varphi - \omega_0^2(3\cos^2\lambda - \sin^2\lambda)\cos^2\varphi]\,.$$

This is the only first integral of (14). No component of the angular momentum is conserved since none of the coordinates is cyclic.

The existence of only one first integral makes the exact solution of (14) a difficult task. Here the problem will be solved in the particular case of small oscillations, which is indeed of great interest in satellite Geodesy.

The pendulum has two stable equilibrium points: $\varphi = 0$, $\lambda = 0$ and $\varphi = 0$, $\lambda = \pi$. Since the system is symmetric with respect to the (x, y) plane it is sufficient to consider the first point only.

Expanding the trigonometric functions in (14) into their power series, and keeping only the first- and second-order terms in the coordinates and velocities, we obtain:

$$\ddot{\varphi} + 3\,\omega_0^2\varphi + 2\,\omega_0\dot{\lambda}\lambda = 0$$

$$\ddot{\lambda} + 4\,\omega_0^2\lambda - 2\,\omega_0\dot{\varphi}\lambda = 0\,. \tag{15}$$

The two equations are coupled by the Coriolis force which is responsible for the terms in $2\,\omega_0$. If the Coriolis force is neglected, the motion in latitude is decoupled from the motion in longitude.

To the second order in φ and λ, the transformations (13) are:

$$x = l\lambda$$

$$y = l\varphi$$

$$z = l\,,$$

from which

$$\varphi = y/l$$

$$\lambda = x/l\,.$$

By this transformation the equations for small oscillations (15) can be written in Cartesian coordinates

$$\ddot{x} + 4\,\omega_0^2 x = (2\,\omega_0/l)x\dot{y}$$

$$\ddot{y} + 3\,\omega_0^2 y = -(2\,\omega_0/l)x\dot{x}\,. \tag{16}$$

The system (16) shows that the coupling introduced by the Coriolis force is a perturbation of the first order in $1/l$.

The non-linear system (16) will be solved approximately by the method of perturbations, i.e.: expanding the solution in series of $1/l$

$$x = x_0 + x_1/l + \dots$$
$$y = y_0 + y_1/l + \dots$$

To the zero order we obtain the unperturbed system

$$\ddot{x} + 4\,\omega_0^2 x = 0$$
$$\ddot{y} + 3\,\omega_0^2 y = 0\,,$$

whose solution is

$$x_0 = A_0 \exp(i2\,\omega_0 t) + B_0 \exp(-i2\,\omega_0 t)$$
$$y_0 = C_0 \exp(i\sqrt{3}\,\omega_0 t) + D_0 \exp(-i\sqrt{3}\,\omega_0 t)\,. \tag{17}$$

The unperturbed trajectory is an open one, since the periods of x and y oscillations are not commensurable. The trajectory fills the rectangle defined by the amplitudes in x and y.

The first-order equation is:

$$\ddot{x}_1 + 4\,\omega_0^2 x_1 = 2\,\omega_0 x_0 \dot{y}_0$$
$$\ddot{y}_1 + 3\,\omega_0^2 y_1 = -2\,\omega_0 x_0 \dot{x}_0\,.$$

Substituting the unperturbed solution (17) into this system we have:

$$\ddot{x}_1 + 4\,\omega_0^2 x_1 = i2\sqrt{3}\,\omega_0^2 [A_0 \exp(i2\,\omega_0 t) + B_0 \exp(-i2\,\omega_0 t)]$$
$$\times [C_0 \exp(i\sqrt{3}\,\omega_0 t) - D_0 \exp(-i\sqrt{3}\,\omega_0 t)]$$
$$\ddot{y}_1 + 3\,\omega_0^2 y_1 = -i4\,\omega_0^2 [A_0^2 \exp(i4\,\omega_0 t) - B_0^2 \exp(-i4\,\omega_0 t)]\,.$$

Solving this non-homogeneous system by means of the Lagrange method, we obtain:

$$x_1 = A_1 \exp(i2\,\omega_0 t) + B_1 \exp(-i2\,\omega_0 t) - i2\{A_0 C_0 \exp[i(2+\sqrt{3})\,\omega_0 t]/(4+\sqrt{3})$$
$$+ A_0 D_0 \exp[i(2-\sqrt{3})\,\omega_0 t]/(4-\sqrt{3}) - B_0 C_0 \exp[-i(2-\sqrt{3})\,\omega_0 t]/(4-\sqrt{3})$$
$$- B_0 D_0 \exp[-i(2+\sqrt{3})\,\omega_0 t]/(4+\sqrt{3})\}$$
$$y_1 = C_1 \exp(i\sqrt{3}\,\omega_0 t) + D_1 \exp(-i\sqrt{3}\,\omega_0 t)$$
$$+ i4/13\,[A_0^2 \exp(i4\,\omega_0 t) - B_0^2 \exp(-i4\,\omega_0 t)]\,.$$

The complex constants A_0, B_0, C_0 and D_0 are determined by the initial conditions which have to be satisfied by the unperturbed solution (17). The constants A_1, B_1, C_1 and D_1 – which are also complex – are determined by the condition that the perturbation in the orbit and the one in the velocity are zero at the initial time.

The reality of the motion implies the following relations between the constants of integration: $B_0 = A_0^*$, $D_0 = C_0^*$, $B_1 = A_1^*$, $D_1 = C_1^*$ (the asterisk indicates the complex conjugate). Then the unperturbed solution and the perturbation can be written more simply:

$$\begin{aligned} x_0 &= a_{x0} \cos(2\,\omega_0 t + \psi_{x0}) \\ y_0 &= a_{y0} \cos(\sqrt{3}\,\omega_0 t + \psi_{y0}) \end{aligned} \tag{18}$$

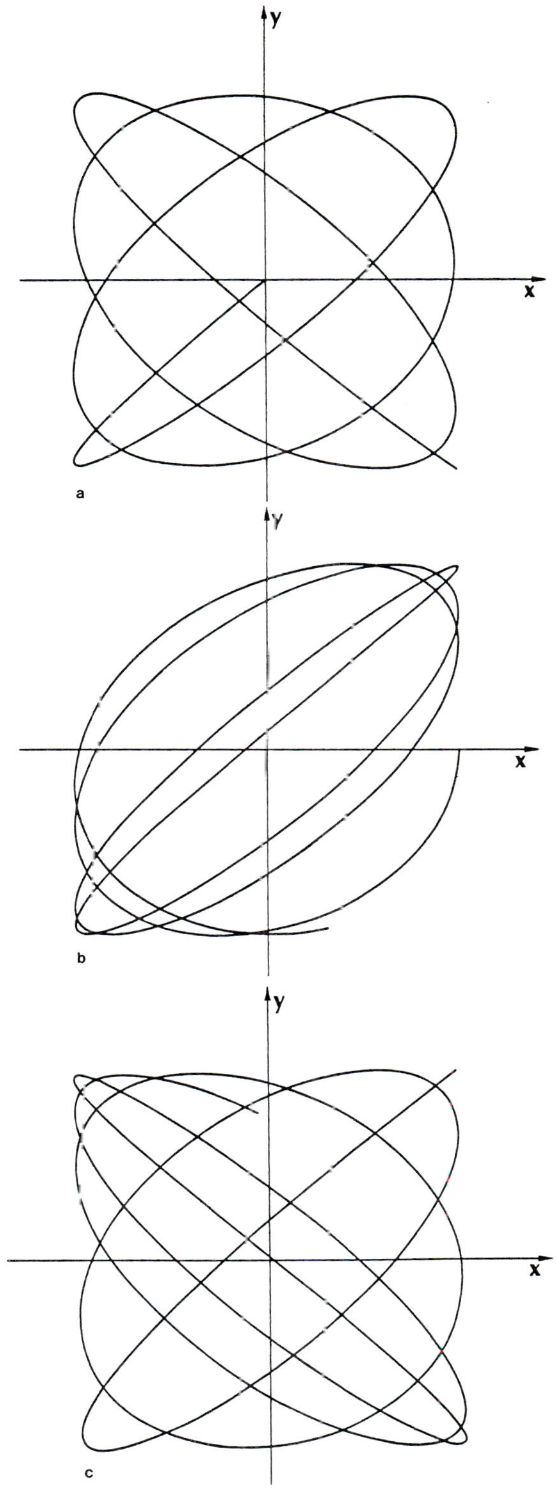

Fig. 4. a – c Some examples of trajectories of the pendulum. The amplitude of the motion is small compared to the length of the pendulum

$$x_1 = a_{x1}\cos(2\omega_0 t + \psi_{x1}) + a_{x0}a_{y0}\{\sin[(2+\sqrt{3})\,\omega_0 t + \psi_{x0} + \psi_{y0}]/(4+\sqrt{3})$$
$$+ \sin[(2-\sqrt{3})\,\omega_0 t + \psi_{x0} - \psi_{y0}]/(4-\sqrt{3})\}$$
$$y_1 = a_{y1}\cos(\sqrt{3}\,\omega_0 t + \psi_{y1}) - 2/13\,a_{x0}^2\sin(4\omega_0 t + 2\psi_{x0}) \tag{19}$$

where all quantities are real and $a_{x0} = 2\,|A_0|$, $a_{y0} = 2\,|C_0|$, $a_{x1} = 2\,|A_1|$, $a_{y1} = 2\,|C_1|$, $\psi_{x0} = \arg(A_0)$, $\psi_{y0} = \arg(C_0)$, $\psi_{x1} = \arg(A_1)$, $\psi_{y1} = \arg(C_1)$.

Some examples of pendulum trajectories are shown in Fig. 4.

The Reaction of the Constraint

The reaction of the constraint can be determined by the following method (Sommerfeld 1949, pp. 103–104). Adding Λx, Λy, Λz in order to the three Cartesian equations (4), then multiplying the new equations by x, y, z in order and summing, one obtains the following equation:

$$\ddot{x}x + \ddot{y}y + \ddot{z}z + \omega_0^2(x^2 - 3z^2) - 2\omega_0(y\dot{z} - \dot{y}z) + \Lambda(x^2 + y^2 + z^2) = 0\,. \tag{20}$$

The first three terms may be written as:

$$\ddot{x}x + \ddot{y}y + \ddot{z}z = \frac{1}{2}\,\frac{d^2}{dt^2}(x^2 + y^2 + z^2) - (\dot{x}^2 + \dot{y}^2 + \dot{z}^2)\,.$$

By the constraint equation (8), the derivative in the right hand side vanishes and (20) reduces to

$$\dot{x}^2 + \dot{y}^2 + \dot{z}^2 - \omega_0^2(x^2 - 3z^2) + 2\omega_0(y\dot{z} - \dot{y}z) - \Lambda l^2 = 0\,.$$

Now the reaction R_N is

$$R_N = \Lambda l = [\dot{x}^2 + \dot{y}^2 + \dot{z}^2 + \omega_0^2(3z^2 - x^2) + 2\omega_0(y\dot{z} - \dot{y}z)]/l\,. \tag{21}$$

R_N is positive towards the centre.

Eliminating z from (21) and the constraint equation (8), expanding the functions of the coordinates x, y and of the velocities $\dot{x}$, $\dot{y}$ into power series and keeping only the first and second order terms, we obtain the expression of the reaction for small oscillations.

$$R_N = l[3\omega_0^2 - 2\omega_0\dot{y}/l + (\dot{x}^2 + \dot{y}^2 - 4\omega_0^2x^2 - 3\omega_0^2y^2)/l^2]\,. \tag{22}$$

The first term – i.e.: $3\omega_0^2 l$ – in the right hand side is the reaction at rest in the equilibrium point $(0, 0, l)$. The second term is due to the Coriolis force and dominates the variable part of the reaction, the other ones being negligible as they are of the second order in $1/l$.

Conclusions

The use of a pendulum attached to a spacecraft – a tethered satellite – has been proposed by O. Colombo to sense the Eötvös tensor – i.e. the second gradient of the Earth's gravitational potential.

The expression of the Eötvös tensor (Marussi, 1979) is:

$$W_{ij} = \begin{vmatrix} -gk_1 & 0 & -gf_1 \\ 0 & -gk_2 & -gf_2 \\ -gf_1 & -gf_2 & gH \end{vmatrix} . \tag{23}$$

In the case studied in this paper it simplifies to

$$W_{ij} = \begin{vmatrix} -gk & 0 & 0 \\ 0 & -gk & 0 \\ 0 & 0 & 2gk \end{vmatrix} .$$

This expression can be used to approximate the Eötvös tensor of the Earth's field.

In spacecraft orbiting in the Earth's field the reaction at rest $3\,\omega_0^2 l$ could be measured to obtain information on the variation of $\omega_0^2 = gk$ along the orbit. The motion of the pendulum produces perturbing variations in the reaction. The magnitude of these perturbations can be estimated by (22).

In expression (22) we can drop the terms in $1/l^2$ and compare the Coriolis term with the reaction at rest. As an example, suppose that the angular amplitude of oscillations is 0.01 rad (about half degree). The amplitude of $\dot{y}$ is approximately $0.01\sqrt{3}\,\omega_0 l$ and the perturbation of R_N is approximately:

$$2\,\omega_0 |\dot{y}|/l = 0.035\,\omega_0^2 .$$

At ground level $\omega_0^2 = 1536\,\text{E}$ ($1\,\text{E} = 1$ eötvös $= 10^{-9}\,\text{s}^{-2}$), at a height of 1000 km above the Earth's surface, $\omega_0^2 = 996$ E, and thus the perturbation in R_N/l is of the order of tens of eötvös: four orders of magnitude larger than the accuracy aimed at by gradiometer measurements (Colombo and Kleusberg 1983). However, measurements of the Eötvös tensor have the purpose to determine the high degree harmonic coefficients of the potential, which induce in the reaction variations of a frequency much higher than the frequency of rotation of the satellite. From the solutions (18) and (19) we see that small oscillations of the pendulum induce reaction variations of low frequency – i.e. of the order of the frequency of rotation. This would allow to filter out the effect of the motion of the pendulum.

Acknowledgement. This study was carried out during the time of Prof. Marussi's illness. He had just the time to revise it almost completely. I am grateful to Prof. Marussi for having proposed me to cooperate in this research (C. Chiaruttini).

Bibliography

Colombo O, Kleusberg A (1983) Applications of an orbiting gravity gradiometer. Bull Gèod 57:83 – 101

Marussi A (1979) The tidal field of a planet and the related intrinsic reference systems. Geophys J R Astr Soc 56:409 – 417

Marussi A (1982) On the structure of the tidal field. Proc Summer School in the Mountains, Admont, Austria

Marussi A (1984) Microgravitation in space. Geophys J R Astr Soc. 76:691 – 695

Sommerfeld A (1957) Lezioni di Fisica Teorica. Vol. I Meccanica. Sansoni, Florence

Appendix
Notations of Vector Analysis – the Vectorial Homographies of Burali-Forti and Marcolongo

W. I. REILLY

1. In his earlier works, and particularly in the *Fondamenti di Geodesia Intrinseca* of 1951, Professor Marussi employs a notation of linear vector operators, founded upon the system of vectorial homographies of Burali-Forti and Marcolongo, which may be unfamiliar to many readers, at least by comparison with the Gibbs/Heaviside notation of the vector calculus, and with the notation of the tensor calculus. The latter was introduced by Professor Marussi himself in the later chapters of the above work, continued in subsequent papers, and formed the basis of a thorough-going presentation of geodetic application in the *Mathematical Geodesy* of Brigadier Hotine (1969).

The following notes may help to elucidate this earlier notation and to illustrate it by setting out a few fundamental relationships both in terms of vectorial homographies and in the notation of the tensor calculus.

2. In developing his system of *Intrinsic Geodesy,* Professor Marussi has sought, as he reiterated so many times, to free Geodesy from the arbitrary and the accidental, including dependence on arbitrary choices of coordinate systems. It is therefore most natural that the should have availed himself of the methods of vector analysis developed by his compatriots, Professor C. Burali-Forti of Turin, and Professor R. Marcolongo of Naples, the outlines of whose system can be found, for instance, in their *Analyse Vectorielle Générale. I: Transformations Linéaires* (Pavia 1912). In the introduction to this work, they propound the advantages of their system, particularly in comparison with those of Hamilton and of Gibbs, especially in terms of its independence of coordinate systems; *inter alia* they remark that "the ordinary vector calculus is a shorthand for Cartesian coordinates" and that "our calculus differs substantially from the ordinary ones, in that it can operate directly on geometric and physical elements, without having any need of recourse to any coordinate. Our calculus can thus be called *intrinsic,* or *absolute,* or *autonomous"* (Burali-Forti and Marcolongo 1912, p. VII). They later continue (p. X): "Differential operators of various orders are elements of the highest importance in the vector calculus. We have derived them all from a single differential operator, that which gives the derivative of an element h with respect to the point P of which h is a function, an operator which we have denoted by $\frac{d}{dP}$ in following the usual notation of Leibniz.

If h is a point, or a vector, $\frac{dh}{dP}$ is a homography whose first invariant and vector double give, respectively, the ordinary divergence and the rotational (or curl) of h. If h is a vectorial homography, $\frac{dh}{dP}$ is an operator which, when applied to a vector, produces a homography".

3. The tensor calculus claims equally to deal with intrinsic or absolute entities, but by contrast it does not seek to conceal the coordinate systems which underlie tensors, indicating by the use of indices the components, with respect to a coordinate system, of the various geometric and physical quantities. Relationships are expressed in terms of these components, but in such a general way that any tensor equation holds in any coordinate system whatsoever. The operation of differentiation is defined in such a way that the resulting (covariant) derivative, with respect to position, is always equivalent to the ordinary Cartesian derivative in a Euclidean space. If the components of a vector are denoted by u_r, ($r = 1, 2, 3$), a tensor of the first rank, then differentiation produces a tensor of the second rank (the equivalent of the homography), denoted by $u_{r,s}$ or $u_{r/s}$ or simply by u_{rs}.

4. In the vector calculus of Gibbs (e.g., Weatherburn 1924), differentiation with respect to position is expressed by the operator $\nabla = \frac{d}{dx}\boldsymbol{i} + \frac{d}{dy}\boldsymbol{j} + \frac{d}{dz}\boldsymbol{k}$, a quasi-vector in which (x, y, z) are the three Cartesian coordinate variables, and $(\boldsymbol{i}, \boldsymbol{j}, \boldsymbol{k})$ the three mutually orthogonal unit vectors in the directions of the coordinate axes.

5. The symbols for the binomial operators for products differ between the different systems. Burali-Forti and Marcolongo use the notation of

$$\boldsymbol{a} \times \boldsymbol{b} \quad \text{and} \quad \boldsymbol{a} \wedge \boldsymbol{b}$$

for the scalar (inner) and vector products of vectors, respectively. The equivalent Gibbsian forms are

$$\boldsymbol{a} \cdot \boldsymbol{b} \quad \text{and} \quad \boldsymbol{a} \times \boldsymbol{b} .$$

In the tensor calculus, the scalar product is denoted by the contracted product of two vectors, in which the same dummy index appears twice, one above and once below, viz.

$$a_r b^r = \sum_{r=1,3} a_r b^r = a_1 b^1 + a_2 b^2 + a_3 b^3 .$$

The vector product is formed by contracting the two vectors with an e-system tensor, or alternating tensor of Ricci, of three indices in a three-dimensional space, giving

$$e^{rst} a_s b_t .$$

6. Returning to the question of linear vector operators, Burali-Forti and Marcolongo define their vectorial homographies as "linear operators which transform vectors into vectors" (Burali-Forti and Marcolongo 1912, p. 11). The definition admits that a real number can be a homography, as it transforms a vector into another vector parallel to the first. Thus if α is a homography, and $\boldsymbol{u}$ a vector, the symbol $\alpha\boldsymbol{u}$ represents a vector. Burali-Forti and Marcolongo further define the homography α as axial if there exists a vector $\boldsymbol{u}$ such that $\alpha = \boldsymbol{u} \wedge$, i.e., the linear vector function $\alpha\boldsymbol{x} = \boldsymbol{u} \wedge \boldsymbol{x}$, is simply a vector product; thus, the resultant vector always lies in the plane perpendicular to $\boldsymbol{u}$, whatever the vector $\boldsymbol{x}$.

However, the most general homography (the "proper homography") is a function of nine independent quantities in a three-dimensional space.

The analogous linear vector operator in the system of Gibbs is the dyadic, of the form

$$\boldsymbol{a}\boldsymbol{i}+\boldsymbol{b}\boldsymbol{j}+\boldsymbol{c}\boldsymbol{k}$$

in which both $(\boldsymbol{a}, \boldsymbol{b}, \boldsymbol{c})$ and $(\boldsymbol{i}, \boldsymbol{j}, \boldsymbol{k})$ are sets of non-coplanar vectors. Pre- and post-multiplication by a vector $\boldsymbol{u}$ leads to the distinct resultant vectors

$$\boldsymbol{u}\cdot(\boldsymbol{a}\boldsymbol{i}+\boldsymbol{b}\boldsymbol{j}+\boldsymbol{c}\boldsymbol{j}) = (\boldsymbol{u}\cdot\boldsymbol{a})\boldsymbol{i}+(\boldsymbol{u}\cdot\boldsymbol{b})\boldsymbol{j}+(\boldsymbol{u}\cdot\boldsymbol{c})\boldsymbol{k}$$

$$(\boldsymbol{a}\boldsymbol{i}+\boldsymbol{b}\boldsymbol{j}+\boldsymbol{c}\boldsymbol{k})\cdot\boldsymbol{u} = \boldsymbol{a}(\boldsymbol{i}\cdot\boldsymbol{u})+\boldsymbol{b}(\boldsymbol{j}\cdot\boldsymbol{u})+\boldsymbol{c}(\boldsymbol{k}\cdot\boldsymbol{u})$$

using the Gibbsian dot-product notation. The dyadic in its general form also has nine independent parameters, for any dyadic is equivalent to three scalars and two sets of orthogonal unit vectors.

In the notation of the tensor calculus, the general linear vector operator is a second-rank tensor of nine independent components, of the form c^{rs}, which by contraction with a vector a_i can yield two distinct resultant vectors

$$c^{rs}a_r \quad \text{and} \quad c^{sr}a_r .$$

7. The dyadic of Gibbs is defined as the sum of three *dyads*, each in the form of the outer product of two vectors, such that a different result is produced by pre- and post-multiplication. Burali-Forti and Marcolongo introduce the dyad as a particular form of homography, with the notation

$$H(\boldsymbol{u},\boldsymbol{v}) = \boldsymbol{v}\boldsymbol{u} \qquad \text{(of Gibbs)}$$

and such that $H(\boldsymbol{u},\boldsymbol{v})\boldsymbol{x} = \boldsymbol{u}\times\boldsymbol{x}\cdot\boldsymbol{v}$ (using here the notation of Burali-Forti and Marcolongo), the resultant vector being parallel to $\boldsymbol{v}$. It should be noted that the homographies of Burali-Forti and Marcolongo are strictly defined as left-operators, and always precede the vectors on which they operate.

In the tensor calculus, the equivalent of the dyad is likewise the second-rank tensor formed by the outer product of two vectors, e.g.,

$$c_{rs} = a_r b_s ,$$

which can, as for any general second-rank tensor, be contracted with a vector u_i in two ways

$$c_{rs}u^r = a_r b_s u^r = (a_r u^r)\, b_s$$

and

$$c_{rs}u^s = a_r b_s u^s = (b_s u^s)\, a_r$$

yielding in the first case a vector parallel to b_s, and in the second a vector parallel to a_r.

8. Burali-Forti and Marcolongo define the homography α as a *dilatation* if for any arbitrary vectors $\boldsymbol{x}$ and $\boldsymbol{y}$

$$\boldsymbol{x}\times\alpha\boldsymbol{y} = \boldsymbol{y}\times\alpha\boldsymbol{x} .$$

For the dyadic of Gibbs, the equivalent is the form which yields the identical resultant vector for both pre- and post-multiplication, e.g., the form $(\boldsymbol{aa}+\boldsymbol{bb}+\boldsymbol{cc})$. In the tensor calculus, the equivalent form is the symmetric tensor of the second rank, $c_{rs}=c_{sr}$. In every case, the number of independent components is reduced from 9 to 6.

9. In view of the restriction of the homography as a left-operator, the device of alternating the dummy index in the tensor calculus, $c^{rs}a_r$ and $c^{sr}a_r$, to produce two distinct resultant vectors, is not available. Burali-Forti and Marcolongo introduce a conjugation operator K, and define the conjugate of the homography α as the homography $K\alpha$, with the same effect as the transposition of indices in a second-rank tensor. For the dyad, $KH(\boldsymbol{u},\boldsymbol{v})=H(\boldsymbol{v},\boldsymbol{u})$. They further define $D\alpha$ as the *dilatation* of the homography α (equivalent to extracting the symmetric part of a second-rank tensor), and V as the *vector* of the homography α. From this follows their statement of the decomposition of a homography (Burali-Forti and Marcolongo 1912, p. 25) "Any homography can always, and in a unique manner, be reduced to the sum of a dilatation and an axial homography". In the adopted notation, this is expressed as

$$\alpha = D\alpha + V\alpha\wedge ,$$

in which $D\alpha$ is the dilatation, and $V\alpha\wedge$ the axial homography. The analogous decomposition is expressed in tensor notation by making use of the properties of the generalized Kronecker delta,

$$\begin{aligned} c_{rs} &= \frac{1}{2}(c_{rs}+c_{sr})+\frac{1}{2}(c_{rs}-c_{sr}) \\ &= \frac{1}{2}(c_{rs}+c_{sr})+\frac{1}{2}\delta^{pq}_{rs}c_{pq} \\ &= \frac{1}{2}(c_{rs}+c_{sr})+\frac{1}{2}e_{jrs}(e^{jpq}c_{pq}) , \end{aligned}$$

from which it is seen that the dilatation $D\alpha$ has its analogue in the tensor expression $\frac{1}{2}(c_{rs}+c_{sr})$ and that the vector $V\alpha$ of the homography α has as its tensor analogue the vector $\frac{1}{2}e^{jpq}c_{pq}$.

10. Burali-Forti and Marcolongo define the first and second scalar invariants of the homography α, denoted by $I_1\alpha$ and $I_2\alpha$ respectively, in terms of the application of α to a general triplet of non-coplanar vectors, or in an equivalent but no less general form, by the introduction of a right-handed orthogonal set of unit vectors $(\boldsymbol{i},\boldsymbol{j},\boldsymbol{k})$, such that

$$I_1 = \boldsymbol{i}\times\alpha\boldsymbol{i}+\boldsymbol{j}\times\alpha\boldsymbol{j}+\boldsymbol{k}\times\alpha\boldsymbol{k}$$

$$I_2 = (\alpha\boldsymbol{j})\wedge(\alpha\boldsymbol{k})\times\boldsymbol{i}+(\alpha\boldsymbol{k})\wedge(\alpha\boldsymbol{i})\times\boldsymbol{j}+(\alpha\boldsymbol{i})\wedge(\alpha\boldsymbol{j})\times\boldsymbol{k} .$$

In the tensor notation, introducing the covariant form g_{rs} and the controvariant form g^{rs} of the metric tensor, the analogue of the first invariant $I_1\alpha$ is

$$g^{rs}c_{rs}$$

and the analogue of the second invariant $I_2\alpha$ is

$$\frac{1}{2} g_{ij} e^{irp} e^{jsq} c_{rs} c_{pq} .$$

11. Burali-Forti and Marcolongo define the *rotation* (or *curl*) and the divergence of a vector in very much the same way as for the Gibbsian notation, but reserve the symbol *grad* for the gradient of an homography; but since a real number is admitted as a special type of homography, then grad as the notation for the gradient of a scalar is also implicitly admitted to their system, and Marussi uses it in this sense. Admitting the scalar W as an homography, the gradient function of Burali-Forti and Marcolongo is

$$\operatorname{grad} W = \left(\frac{dW}{dP}\boldsymbol{i}\right)\boldsymbol{i} + \left(\frac{dW}{dP}\boldsymbol{j}\right)\boldsymbol{j} + \left(\frac{dW}{dP}\boldsymbol{k}\right)\boldsymbol{k}$$

while in Gibbsian notation $\operatorname{grad} W = \nabla W$, and the analogue in the tensor notation is simply the covariant derivative $W_{,r}$ $(= W_{/r})$.

The *rotation* (*curl*) and *divergence* of a vector $\boldsymbol{u}$ are given in the same three systems by:

$$\operatorname{rot}\boldsymbol{u} = 2V\frac{d\boldsymbol{u}}{dP} \qquad \nabla \times \boldsymbol{u} \qquad e^{rst} u_{s,t}$$

$$\operatorname{div}\boldsymbol{u} = I_1\frac{d\boldsymbol{u}}{dP} \qquad \nabla \cdot \boldsymbol{u} \qquad g^{st} u_{s,t} .$$

12. Having briefly sketched the notation, we shall turn to some concrete examples of Marussi's usage in the *Fondamenti di Geodesia Intrinseca.* Marussi (1951): His equation (1.1) introduces the gravity vector $\boldsymbol{g}$, firstly as the gradient of the scalar potential function W, and secondly as the product of the intensity of gravity g and a unit vector (or versor) $\boldsymbol{n}$ in the zenith direction, viz.

loc. cit. (1.1) $\quad \boldsymbol{g} = \operatorname{grad} W = -g\boldsymbol{n}$.

He then introduces two homographies in the sense of gradients of vectors, the first being the gradient of the gravity vector $\boldsymbol{g}$,

$\mathscr{W} = \dfrac{d\boldsymbol{g}}{dP}$ and named the homography of Eötvös, and the second, the gradient of the zenith vector $\boldsymbol{n}$,

$\sigma = \dfrac{d\boldsymbol{n}}{dP}$ and named the generalized homography of Burali-Forti.

This leads to the equation expressing the gradient of Eq. (1.1) as

loc. cit. (1.5) $\quad \mathscr{W} = -g\sigma - H(\operatorname{grad} g, \boldsymbol{n})$ introducing the symbol for the dyad.

The tensor analogues of these two equations are

$$W_r = -g n_r ,$$

where W_r is the gravity vector and n_r the zenith vector; and

$$W_{r,s} = -g n_{r,s} - g_s n_r ,$$

in which $W_{r,s}(= W_{rs})$ is the "homography of Eötvös", $n_{r,s}(= n_{rs})$, the gradient of the zenith vector, is the "generalized homography of Burali-Forti", and g_s is the gradient of the intensity of gravity. For the latter, Marussi (1951) introduces

loc. cit. (1.10) $\quad \boldsymbol{G} = \operatorname{grad} g = -\mathscr{W}\boldsymbol{n}$.

The curvature of the vertical, equal to the surface gradient of gravity, is given by Marussi (1951) as

loc. cit. (4.3) $\quad \sigma\boldsymbol{n} = \frac{1}{g}\operatorname{grad}_{\Sigma} g$

and in tensor notation as $n_{r,s}n^s$.

The expression

loc. cit. (1.14) $\quad K\sigma\boldsymbol{n} = 0$

and the tensor analogue $n_{r,s}n^r = 0$ express the result that the gradient of a unit vectors has no component parallel to that vector.

Subsequent discussion of the intrinsic properties of the gravity field is largely developed in terms of the homographies $\mathscr{W}$ and σ, the vectors $\boldsymbol{G}$ and $\boldsymbol{n}$, and the scalar g.

In Marussi's later works, as characterized, for example, by his Erice lectures of 1974, the vectorial homographies of Burali-Forti and Marcolongo have ceased to appear explicitly and have been largely replaced by the notation of the tensor calculus, but their notation for scalar and vector products is still used, and the influence of their concept of an intrinsic or absolute calculus is never far from sight.

Bibliography

Burali-Forti C, Marcolongo R (1912) Analyse vectorielle générale. Transformations linéaires. Mattei, Pavia

Hotine M (1969) Mathematical geodesy. US Dep Commerce, ESSA Monogr 2

Marussi A (1951) Fondamenti di Geodesia intrinseca. Pubbl Comm Geodet Ital Mem 7 (Chap. I.2, this Vol.)

Weatherburn CE (1924) Advanced vector analysis. Bell, London

Bibliography

of Prof. ANTONIO MARUSSI

Marussi, A., 1938. Sul ripristino di un punto trigonometrico scomparso. L'Universo, 8, XIX.

Marussi, A., 1940. Condizioni di optimum per la forma dei triangoli geodetici. L'Universo, 2.

Marussi, A., 1941. Il Paleotimavo e l'antica idrografia subaerea del Carso Triestino. Boll. Soc. Adriatica di Sc. Naturali, XXXVIII.

Marussi, A., 1941. Ipotesi sullo sviluppo del Carsismo. Giornale di Geologia, serie II, XV.

Marussi, A., 1941. Ricerca dei coefficienti di correlazione fra osservazioni: applicazione a problemi di geodesia. Rivista del Catasto e SS. TT. EE., XIX, 4.

Marussi, A., 1942. Tavola per le riduzioni normali di osservazioni azimutali all'ellissoide di riferimento ed all'arco di geodetica – calcolata di grado in grado da 35 a 48 gradi di latitudine. Boll. Geod. de: l'Universo, XXIII, 9.

Marussi, A., 1944. Sulla livellazione trigonometrica geoidica. L'Universo, XXV, 5.

Marussi, A., 1946. Le triangolazioni eseguite dall'Italia e dall'Austria in Alto Adige, e la loro reciproca dipendenza. Boll. Geod. de: L'Universo, XXVI, 3.

Marussi, A., 1946. Saggio storico di cartografia giuliana dai primordi al secolo XVIII. Club Alpino Italiano. Soc. Alpina delle Giulie, Trieste.

Marussi, A., 1949. Fondements de géométrie différentielle absolue du champ potentiel terrestre. Bulletin Géodésique, 14.

Marussi, A., 1949. Geodesia intrinseca. Università di Trieste.

Marussi, A., 1949. La nuova triangulazione di I e II ordine della Sicilia. Boll. Geod. e Sc. Affini, I.G.M., VIII, 2.

Marussi, A., 1950. Tre conversazioni radiotrasmesse per l'Università Internazionale Marconi:
- La tradizione cartografica italiana.
- Come si crea una carta topografica moderna.
- Nuovi orizzonti per la Geodesia.

Marussi, A., 1950. Principi di Geodesia intrinseca applicati al campo di Somigliana. Boll. Geod. e Sc. Affini, I.G.M., IX, 2.

Marussi, A., 1950. Sviluppi di Legendre generalizzati per una curva qualunque tracciata su di una superficie pure qualunque. Rend. Acc. Naz. Lincei, Cl. Sc. Mat. Fis. Nat., Serie VIII, VIII, 4.

Marussi, A., 1950. Sviluppi di Legendre generalizzati nello spazio. Rend. Acc. Naz. Lincei. Cl. Sc. Mat. Fis. Nat., Serie VIII, VIII, 5.

Marussi, A., 1950. Sviluppi di Legendre generalizzati per una curva qualunque dello spazio. Rend. Acc. Naz. Lincei, Cl. Sc. Mat. Fis. Nat., Serie VIII, IX, 1–2.

Marussi, A., 1950. Sulla variazione con l'altezza dei raggi di curvatura principali nella teoria di Somigliana. Boll. Geod. e Sc. Affini, I.G.M., IX, 1.
Marussi, A., 1950. La teoria della proiezione di Gauss con premesse di geometria differenziale. Lezioni tenute al XII corso di Geodesia dell'I.G.M. nell'anno accademico 1950.
Marussi, A., 1950. Il primo problema fondamentale della geodesia ampliato nel campo di Somigliana. Riv. di Geof. pura e applicata, XVIII, Volume dedicato al Prof. Somigliana.
Marussi, A., 1950 – 51. Sul calcolo dei simboli di Christoffel per le proiezioni di Mercatore e di Gauss. Atti dell'Ist. Veneto di Sc. Lettere ed Arti, CIX.
Marussi, A., 1951. Fondamenti di Geodesia intrinseca. Comm. Geod. It., "Memorie", 7.
Marussi, A., 1951. Su alcune proprietà integrali delle rappresentazioni conformi di superfici su superfici. Rend. Acc. Naz. Lincei, Cl. Sc. Fis. Mat. Nat., Serie VIII, X, 4.
Marussi, A., 1951. Determinazione a priori del modulo di deformazione lineare nella rappresentazione conforme di Gauss. Rend. Acc. Naz. Lincei. Cl. Sc. Fis. Mat. Nat., Serie VIII, XI, 3 – 4.
Marussi, A., 1951. Les principes de la Géodésie intrinsèque. Bull. Géod. de l'A.I.G., 19.
Marussi, A., 1951. Sulla rappresentazione del geoide sull'ellissoide. Boll. Geod. e Sc. Affini, I.G.M., X, 3.
Marussi, A., 1951. Su alcune proprietà fondamentali delle rappresentazioni conformi fra superfici. Boll. Geod. e Sc. Affini, I.G.M., X, 2.
Marussi, A., 1952. Determinazione dell'angolo fra la tangente in un estremo e la corrispondente sezione normale contenente l'altro estremo, per un arco finito di geodetica su di una superficie qualunque. Rend. Acc. Naz. Lincei, Cl. Sc. Fis. Mat. Nat., Serie VIII, XII, 5.
Marussi, A., 1952. Sulla curvatura e torsione del campo di gravità. Annali di Geofisica, V, 2.
Marussi, A., 1951 – 52. Le coordinate intrinseche della Geodesia. Rend. Seminario Matematico dell'Univ. e Politecnico di Torino, 11.
Marussi, A., 1952. Generalizzazione del teorema di Dalby per una superficie qualunque. Festschrift Eduard Dolezal.
Marussi, A., 1953. Un'analogia fra le leggi della propagazione della luce in mezzi rifrangenti continui isotropi, e le rappresentazioni conformi. Atti Fondaz. G. Ronchi dell'Ist. Naz. Ottica, VIII, 1.
Marussi, A., 1953. La torsione di una traiettoria luminosa in un mezzo rifrangente continuo isotropo. Atti Fondaz. G. Ronchi dell'Ist. Naz. Ottica, VIII, 2.
Marussi, A., 1953. Sulla riduzione regionale e sul metodo delle derivate seconde in gravimetria. Annali di Geofisica, VI, 2.
Marussi, A., 1953. Sulla divergenza del campo gravimetrico normale. Geofisica Pura e Applicata, 25.
Marussi, A., 1953. Sul significato del secondo parametro differenziale della gravità in Geofisica. Geofisica Applicata, 2.
Marussi, A., 1953 – 54. Sui parametri differenziali del campo gravimetrico. Rend. del Seminario Matem. e Fis. di Milano, XXV.

Marussi, A., 1954. Sulla curvatura tangenziale delle trasformate di curve nelle rappresentazioni affini fra superfici. Rend. Acc. Naz. Lincei, Cl. Sc. Fis. Mat. Nat., Serie VIII, XVI, 4.

Marussi, A., 1953–54. Osservazioni gravimetriche e geomagnetiche effettuate dalla Spedizione Italiana al Karakorum, 1954. Boll. della Soc. Adr. di Sc. Nat., Trieste, 42.

Marussi, A., 1954–55. L'esplorazione geofisica del Karakorum. Discorso inaugurale per l'anno accademico 1954–55 dell'Università di Trieste.

Marussi, A., 1955. Lo stato attuale dell'esplorazione geofisica del Karakorum. Geofisica e Meteorologia, III, 1–2.

Desio, A., Marussi, A. and Caputo, M., 1955. Glaciological research of the Italian Karakorum Expedition 1953–55. Publ. n. 54, of the I.A.S.H.

Marussi, A., 1955. Con la Spedizione Italiana al K2. L'esplorazione geofisica del Karakorum, 1955. Publicaciones de la Real Sociedad Geogràfica, Serie B, 355.

Marussi, A., 1956. Expediçao Italiano ao Karakorum. Separata do Buletin de Sociedade de Geografia de Lisboa, Julho-Setembre.

Marussi, A., 1956. Gravimetric and magnetometric surveys performed by the Italian Karakorum Expedition 1954–55. Bull. Géod. de l'A.I.G., 41.

Marussi, A., 1957. Sulle rappresentazioni fra superfici definite mediante la forma quadratica che ne determina il modulo di deformazione. Festschrift zum 75. Geburtstag von C. F. Baeschlin, Zürich.

Marussi, A., 1957. La gravità nel Karakorum nel Turkestan Cinese e nel Pamir. Geofisica e Meteorologia, V, 3–4.

Marussi, A., 1959. Les résultats des recherches géophysiques de l'expédition Italienne au Karakorum en 1954–55. Ciel et Terre, LXXV, 7–8.

Marussi, A., 1959. Dalla geodesia classica alla geodesia in tre dimensioni. Boll. Geod. e Sc. Affini, I.G.M., XVIII, 4.

Marussi, A., 1959. Une nouvelle position dans la théorie des cartes géographiques. Communication de l'Observatoire Royal de Belgique, 192, Série Géophysique, 49.

Marussi, A., 1959. The University of Trieste Station for the study of tides of the vertical in the Grotta Gigante. Proceedings of the Third Intern. Symposium on Earth Tides, Trieste.

Marussi, A., 1960. Wo steht die Geodäsie heute? Mitteilungen aus dem Geodätischen Institut Potsdam Nr. 39, Sonderdruck aus der Zeitschrift "Vermessungstechnik", 11.

Marussi, A., 1960. I primi risultati ottenuti nella stazione per lo studio delle maree della verticale della Grotta Gigante. Boll. Geod. e Sc. Affini, I.G.M., XIX, 4.

Desio, A. and Marussi, A., 1960. On the geotectonics of the granites in the Karakorum and Hindu Kush Ranges (Central Asia). Int. Geological Congress, Copenhagen.

Marussi, A., 1961. Geodetic and topographic work of the Italian Karakorum Expedition 1953–54. Boll. Geod. e Sc. Affini, I.G.M., XX, 1.

Marussi, A., 1961. Intrinsic coordinates in practical geodesy. The Earth Today. Geophys. J. R. astr. Soc., vol. 4.

Bolt, B. A. and Marussi, A., 1962. Eigenvibrations of the Earth observed at Trieste. Geophys. J. R. astr. Soc., vol. 6, 3.

Marussi, A., 1962. Intrinsic geodesy and earth satellites. II Symposium on geodesy in three dimensions, Cortina d'Ampezzo, May, 1962.
Marussi, A., 1962. The use of artificial satellites for geodesy. Proceedings of the first international symposium on the use of artificial satellites for geodesy, Washington D.C., April 26 – 28. Edited by G. Veis.
Marussi, A., 1963. Le anomalie della gravità lungo la catena del Karakorum-Hindu Kush. Rend. Acc. Naz. Lincei, Cl. Sc. Fis. Mat. Nat., Serie VIII, XXXV, 5.
Abetti, G., Caloi, P. and Marussi, A., 1963. Italian pioneers in the physics of the universe. Cahiers d'Histoire Mondiale, VII, 2.
Marussi, A., 1964. Descrizione sommaria dell'ambiente geografico e delle stazioni di raccolta del materiale zoologico. Atti del Museo Civico di Storia Naturale, Trieste, XXIV, 4.
Marussi, A., 1964. Geophysics of the Karakorum. It. Exp. to the K2 and Hindu Kush, Scient. Rep. II Geophysics, vol. I, 242, E. J. Brill, Leiden.
Cook, A., Marussi, A. and Rowley, W. R. C., 1965. A laser strain gauge for seismology. Geophys. J. R. astr. Soc., vol. 9, 2, 3.
Marussi, A., 1965. Funzioni di dissipazione delle auto-oscillazioni della Terra eccitate dal terremoto del Cile e registrate a Trieste. Rend. Acc. Naz. Lincei, Cl. Sc. Mat. Fis. Nat., Serie VIII, XXXVIII, 3.
Marussi, A., 1965. Impiego dei geofoni nella ricerca di acque carsiche scorrenti. Atti Convegno sul Problema della Acque in Italia. Parte I: le acque sotterranee. Milano, Federaz. delle Ass. Scientif. e Tecniche, Ottobre 1965.
Marussi, A., 1965. Correlazione fra Carsismo epigeo ed ipogeo. Atti Convegno sul Problema delle Acque in Italia. Parte I: le acque sotterranee. Milano, Federaz. delle Ass. Scientif. e Tecniche, Ottobre 1965.
Marussi, A., 1965. Relazione generale sui compiti della Geodesia e Geofisica e loro organizzazione sul piano internazionale. Suppl. al "Nuovo Cimento", 4, serie I, vol. 3.
Marussi, A., 1967. Einige Bemerkungen über die Anwendung der Konformen Abbildungen in der dreidimensionalen Geodäsie. „Aus der Geodätischen Lehre und Forschung" – Festschrift zum 70. Geburtstag von Walter Grossmann, 5. 1. 1967 Konrad Wittwer, Stuttgart.
Marussi, A., 1967. Séries statistiques ordonnées et stationaires en Géodésie. Boll. Geof. Teor. Appl., IX, 33.
Bozzi-Zadro, M. and Marussi A., 1967. Polarization and total energy spectra of the eigenvibrations of the Earth recorded at Trieste. Geophys. J. R. astr. Soc., vol. 12, 4.
Marussi, A., 1967. Ancora sulla funzione di dissipazione delle auto-oscillazioni della Terra eccitate dal terremoto del Cile e registrate a Trieste. Rend. Acc. Naz. Lincei, Cl. Sc. Fis. Mat. Nat., Serie VIII, XLII, 1.
Marussi, A., 1968. Gravity field, inner geometry of. International Dictionary of Geophysics. Pergamon Press.
Marussi, A., 1969. In memory of Martin Hotine. Survey Review, XX, 152.
Marussi, A., 1971. Le oscillazioni libre della Terra. Acc. Naz. Lincei, Quaderno 151.
Marussi, A., 1971. The Trieste astrographic satellite camera. 50 Jahre Wild Heerbrugg 1921 – 1971. Festschrift Geodäsie und Photogrammetrie.

Marussi, A., 1971. Presidential address. International association of geodesy. XVth General Assembly. Bull. Géod., A.I.G., 102.

Marussi, A., 1972. Ambiente e calamità naturali. Nuova Antologia, 2059.

Bozzi-Zadro, M. and Marussi, A., 1973. On the static effect of Moon and Sun on the shape of the Earth. Presented at the 5th Symposium on mathematical geodesy (2nd Hotine Symposium). Florence, 25 – 26 October 1972. Boll. Geod. e Sc. Affini, I.G.M., XXXII, 4.

Marussi, A., 1973. Spettroscopia della Terra. Estratto da: Enciclopedia della Scienza della Tecnica Mondadori, Annuario dell'EST: S&T/73, n. 9, 179 – 188.

Marussi, A., 1973. On the three-dimensional computation of geodetic networks and anomalies. Rend. Acc. Naz. Lincei, Cl. Sc. Mat. Fis. Nat., Serie VIII, LV, 5.

Marussi, A., 1973. L'avvento della Cartografia regionale geodetica. Boll. dell'Associazione Italiana di Cargografia, 28 – 29.

Marussi, A., 1974. Geodetic networks in space. The Canadian Surveyor, 28, 5.

Marussi, A., 1974. Africa and modern geodesy. First Symposium on Geodesy in Africa, Khartoum, Sudan. January 12 – 19, 1974, African Geodetic Journal, vol. 1, 2.

Marussi, A., 1974. On the representation of the actual gravity field of the Earth on the normal ellipsoidal field. Geophys. J. R. astr. Soc., vol. 37, 3.

Marussi, A., Moritz, H., Rapp, R. H. and Vicente, R. O., 1974. Ellipsoidal density models and hydrostatic equilibrium. Interim Report. Phys. Earth Planet. Inter., IX, 1.

Marussi, A., 1975. Geomorphology, paleohydrography and karstification in the karst of Trieste and Upper Istria. Sonderdruck aus „Steir. Beiträge zur Hydrogeologie", 27.

Marussi, A., 1976. Viaggio nella Repubblica Popolare Cinese con la delagazione scientifica italiana. Relazione svolta nella seduta dell'11 giugno 1975. Quaderno n. 220 dell'Acc. Naz. Lincei, anno CCCLXXIII.

Marussi, A., 1976. Intrinsic geodesy elements of tensor calculus. Part 1 and 2. Sonderdruck aus „Methoden und Verfahren der Mathematischen Physik. Band 13. Published also in Boll. Geod. e Sc. Affini, I.G.M., XXXV, 1.

Marussi, A., 1976. Geotettonica delle zone orogeniche del Kashmir Himalaya-Karakorum-Hindu Kush-Pamir. Roma, 25 – 27 giugno 1974. Atti dei Convegni Lincei, n. 21. Roma, 1976. Articles: Introductory Report on Geophysics Gravity in Karakorum.

Marussi, A., 1977. Geodäsie: Zwischen Wirklichkeit und Abstraktion. Österr. Zeitschrift für Vermessungswesen und Photogrammetrie, Jg. 65, 3 – 4.

Marussi, A., 1979. The tidal field of a planet and the related intrinsic reference systems. Geophys. J. R. astr. Soc., vol. 56, 3.

Marussi, A., 1979. Natural reference systems and their reciprocals in geodesy. Finnish Geodetic Inst., Helsinki, 89.

Beloussov, V. V., Belyaevsky, N. A., Borisov, A. A., Volvovsky, B. S., Volvovsky, I. S., Resvoy, D. P., Tal-Virsky, B. B., Khamrabaev, I. Kh., Kaila, K. L., Narain, H., Marussi, A. and Finetti, I., 1980. Structure of the lithosphere along the deep seismic sounding profile: Tien Shan-Pamirs-Karakorum-Himalayas. Tectonophysics, vol. 70, 3 – 4.

Marussi, A., 1980. On the density distribution in bodies of assigned outer Newtonian attraction. Boll. Geof. Teor. Appl., XXII, 86.

Marussi, A., 1980. Differential parameters in the normal gravity field. Rend. Acc. Naz. Lincei, Cl. Sc. Mat. Fis. Nat., Serie VIII, LXIX, 6.

Marussi, A., 1982. The adjustment of geodetic networks in the three-dimensional ellipsoidal model space of Somigliana. Symp. on geodetic networks and computations of the I.A.G., München, 1981, DGK, B, Nr. 258/VI.

Marussi, A., 1982. On the invariants of the gravity field. (Feestbundel ter gelegenheid van de 65ste verjaardag van Prof. Baarda.) Vol. I: Edit. by Techn. Hogeschool – Delft.

Marussi, A., 1982. On the structure of the tidal field. Mitteilungen der Geodätischen Inst. der Technischen Univ. Graz. 41.

Marussi, A., 1982. Geodesy and geotectonics. Mitteilungen der Geodätischen Inst. der Technischen Univ. Graz 41. Boll. Geod. e Sc. Affini, I.G.M., XLII, 3, 1983.

Ebblin, C., Marussi, A., Poretti, G., Rahim, S. M. and Richardus, P., 1983. Gravity measurements in the Karakorum. Boll. Geof. Teor. Appl., XXV, 99–100.

Marussi, A., 1983. On the structure of the microgravitational field of a planet (The Tidal Field). Boll. Geod. e Sc. Affini, I.G.M., XLII, 4.

Marussi, A., 1983. The Pakistani-Italian Karakorum geophysical project in the frame of the Pamir-Himalaya international project. Boll. Geof. Teor. Appl., XXV, 99–100.

Marussi, A., 1983. Geophysical trends and evolution of the Pamirs syntaxis and Karakorum. Boll. Geof. Teor. Appl., XXV, 99–100.

Marussi, A., 1984. Microgravitation in space. Geophys. J. R. astr. Soc., vol. 76, 3.

Marussi, A., 1984. Gravity anomalies and gravity of Karakorum. Inst. of Geology, Punjab Univ., Lahore/Pakistan.

Marussi, A., Slejko, D., Zanolla, C., 1984. Contouring in geographical coordinates. Boll. Geof. Teor. Appl., XXVI, 103.

Marussi, A., Longo, J., 1984. Il progetto Pamir-Himalaya. Le Scienze.

Marussi, A., Chiaruttini, C. The motion of a free particle and of a spherical pendulum in the microgravitational field of a gravitationally stabilized satellite in circular orbit in a central field. In press.

Marussi, A., Sleijko, D., Zanolla, C., 1984. Contouring in Geographical Coordinates. Boll. Geol. Teor. Appl. XXVI, 103.

Index of Abbreviations

(Institutions and Publications)

H. HORNIK

Abh K Bayer Akad Wiss: Abhandlungen der Königlichen Bayerischen Akademie der Wissenschaften, München

AIG: Association Internationale de Géodésie (cf. IAG)

Ann Geofis: Annali di Geofisica

Ann Mat: Annali di Matematica

Ann Mat Pura Appl: Annali di Matematica pura ed applicata

Astron Nachr: Astronomische Nachrichten

Atti Accad Sci Nat: Atti dell'Accademia di Scienze Naturali, Catania

Atti Accad Torino: Atti dell'Accademia di Torino

Atti Acc Naz Lincei: Atti dell'Accademia Nazionale dei Lincei

Atti Fond Giorgio Ronchi: Atti della Fondazione Giorgio Ronchi

Atti Ist Veneto Sci Lett Art: Atti dell'Istituto Veneto delle Scienze, Lettere ed Arti

Atti Mem Accad Virg Mantova: Atti e memorie dell'Accademia Virgiliana di Mantova

Atti R Accad Sci: Atti della Regia Accademia delle Scienze, Bologna

Atti R Accad Sci Lett Belle Art Palermo: Atti Regia Accademia di Scienze, Lettere e Belle Arti di Palermo

Atti R Accad Torino: Atti Regia Accademia di Torino

Atti R Ist Veneto: Atti del Regio Istituto Veneto

Boll Com Ital Geodes Geofis: Bollettino della Commissione Italiana di Geodesia e Geofisica

Boll Geodes Sci Affini: Bollettino di Geodesia e Scienze Affini

Boll Un Mat Ital: Bollettino dell'Unione Matematica Italiana

Bull Astron: Bulletin Astronomique, Paris

Bull Géod: Bulletin Géodésique de l'Association Internationale de Géodésie

Bull Géodés: cf. Bull. Géod.

Can Surv: Canadian Surveyor

Comm Geodet Ital Mem: Commissione Geodetica Italiana – Memorie –

C R Comm Géodés Balt: Comptes Rendues de la Commission Géodésique Baltique

Crelle's J: Journal für die reine und angewandte Mathematik, edited by A. L. Crelle, Berlin

CR XV Conf Gen Assoc Géodés Int: Comptes Rendues du XV Conférence Générale de l'Association Géodésique Internationale

Den Dan Graadmaaling: Den Danske Graadmaaling, København

D G K: Veröffentlichungen der Deutschen Geodätischen Kommission bei der Bayerischen Akademie der Wissenschaften, München

G Battaglini: Giornale di Battaglini, Lit. A. Longo, Palermo

Geofis Pura Appl: Geofisica Pura e Applicata

Geophys J: Geophysical Journal

Geophys J R Astron Soc London: (cf. GJRAS)

Gerlands Beitr Geophy: Gerlands Beiträge zur Geophysik

G J R A S: Geophysical Journal of the Royal Astronomical Society, London

G Sci Mat Econ Palermo: Giornale di Scienze Matematiche ed Economiche di Palermo

G Soc Lett Convers Sci Genova: Giornale della Società di Letture e Conversazioni Scientifiche di Genova

I A G: International Association for Geodesy (cf. AIG)

I G M: Istituto Geografico Militare, Firenze

I G N, Bruxelles: Institut Géographique National, Bruxelles

I G N, Paris: Institut Géographique National, Paris

Inst Géogr Nat: cf. IGN

Inst Géol Pol Warszawa: Institut Géologique de Pologne, Warszawa

Ist Geodes Finn: Publicazioni dell'Istituto di Geodesia Finlandese – Veröffentlichungen des Finnischen Geodätischen Instituts – Publications of the Finnish Geodetic Institute

I U G G: International Union for Geodesy and Geophysics (cf. UGGI)

J Geophys Res: Journal of Geophysical Research

J Math Pure Appl: Journal de mathématique pure et appliquée

J Reine Angew Math: Journal für Reine und Angewandte Mathematik

K Preuss Geodät Inst: Königlich Preußisches Geodätisches Institut, Potsdam

Math Abh Akad Wiss, Berlin: Mathematische Abhandlungen der Königlichen Akademie der Wissenschaften zu Berlin
in: Abhandlungen der Königlichen Akademie der Wissenschaften zu Berlin

Math Phys Pap: Mathematical and Physical Papers, Cambridge

Mem Soc Astron Ital: Memorie della Società Astronomica Italiana

Mitt K u K Milit Geogr Inst, Wien: Mitteilungen des Kaiserlichen und Königlichen Militärgeographischen Instituts, Wien

Mon Not R Astron Soc Geophys: Monthly Notices of the Royal Astronomical Society, London, Geophysical Supplement

Nor Mat Tidensskr: Norsk Matematisk Tidsskrift, Oslo (from 1953: Nordisk ...)

Not Mem Mat: Note e memorie matematiche

Oesterr Akad Sitzungsber Kl II: (Österreichische) Akademie der Wissenschaften in Wien, Mathematisch-Naturwiss. Klasse, Sitzungsberichte, Abteilung II

Osserv Astron Palermo: Osservatorio Astronomico di Palermo

Period Mat: Periodico di Matematica

Philos Trans R Soc: Philosophical Transactions of the Royal Society, London

Phys Reichsanst Wiss Abh (= Wiss Abh Phy-Tech Reichs Anst): Wissenschaftliche Abhandlungen der Physikalisch-Technischen Reichsanstalt, Berlin

Proc R Soc London: Proceedings of the Royal Society, London

Pubbl Ist Sup Ing Milan: Pubblicazioni dell'Istituto Superiore d'Ingegneria di Milano

Pubbl Ist Topogr Geodes R Polit Milano: Pubblicazioni dell'Istituto di Topografia e Geodesia del Regio Politecnico di Milano (Politec)

R Comm Geod Ital: Regia Commissione Geodetica Italiano

Rend Accad Ital: Rendiconti dell'Accademia Italiana

Rend Accad Naz Lincei: Rendiconti dell'Accademia Nazionale dei Lincei

Rend Circ Mat Palermo: Rendiconti del Circolo Matematico di Palermo

Rend Ist Lomb: Rendiconti dell'Istituto Lombardo

Rend R Ist Lomb Sci Lett: Rendiconti del Regio Istituto Lombardo di Scienze e Lettere

Rev Gen Sci Pure Appl: Revue générale des sciences pures et appliquées, Paris

R Osserv Astron Padova: Regio Osservatorio Astronomico di Padova

Sitzungsber K Preuss Akad: Sitzungsberichte der Königlichen Preußischen Akademie der Wissenschaften, Berlin

Trans Cambridge Philos Soc: Transactions of the Cambridge Philosophical Society

Trav Mém Bur Int Poids Mesures: Travaux et Mémoires du Bureau International de Poids et Mesures

U G G I: Union de Géodésie e Géophysique International (cf. IUGG)

Verhandl der internat Erdmessung: Verhandlungen der internationalen Erdmessung, Berlin

Veröff d Preuß Geod Inst, Berlin: Veröffentlichungen des Preußischen Geodätischen Instituts, Berlin

Z Geophys: Zeitschrift für Geophysik

Z Instrumentenkd: Zeitschrift für Instrumentenkunde

Antonio Marussi 1908 – 1984

Antonio Marussi was born on October 12, 1908 in Trieste. Throughout his life, he loved Trieste and was proud of this city, with its special geographical and cultural situation, with its rich ethnic and historical heritage. His father's name, Gustavo Marussig, still had the Friulian ending "ig", and his mother's maiden name was Maria Leutheuser. His uncle, Piero Marussig, was a well-known painter and one of the founders of the school of the Novecento Italiano. The family setting would be entirely incomplete without giving the fully appropriate and highly deserved place of honor to his wife Dolores ("Lori" to her friends), née de Finetti, and the mathematical reader will recall that her brother, Bruno de Finetti, is one of the most outstanding representatives (and critics) of the modern theory of probability.

Marussi's career was straightforward. In 1932 he obtained his doctor's degree in pure mathematics from the University of Bologna. But he was always drawn to applications: after a brief start in 1932 at the Institute of Geophysics of Trieste, in the same year he entered the Military Geographic Institute in Florence where, apart from a 3-years' interruption when he worked as an actuary in Assicurazioni Generali in Trieste and Venice, he was a Geographical Engineer until 1952. In this function he performed practical geodetic and astronomic work not only in Italy, but also in Ethiopia, Albania, and Greece.

Much as he liked his practical work, he could not but yield simultaneously to his penchant for theoretical research, which led him to become Privatdocent and then, in 1952, Professor of Geodesy and Geophysics at the University of Trieste. He founded the Institute of Geodesy and Geophysics of this university and was its director for many years. From 1967 to 1973 he was Dean of the Faculty of Science of his university.

It is not an overstatement to say that modern theoretical geodesy was founded by two men: in 1945, M. S. Molodensky published the paper *Investigations of the Fundamental Theoretical Problems of Geodetic Gravimetry* in Moscow, and in 1949, Antonio Marussi's article *Fondements de Géométrie Différentielle Absolue du Champ Potentiel Terrestre* appeared in the *Bulletin Géodésique,* after he had presented his ideas in 1948 at the first post-war General Assembly of the International Association of Geodesy in Oslo.

These two pioneering papers introduced what is now known as the global and the local treatment of the earth's gravity field. Molodensky regarded the problem of determining the earth's surface from gravimetric data as a boundary-value problem of potential theory, and Marussi initiated the treatment of the gravity field by modern differential geometry.

The present book is an elaborate witness of Marussi's fundamental ideas in the field of *Intrinsic Geodesy* with its interplay of geometry and physics in the study of the Earth's gravitational field. It should be mentioned also that Marussi has influenced modern geodetic thinking not only through his research and his publications, but also by arranging a series of International Symposia on Mathematical Geodesy in Italy (Venice, 1959; Cortina d'Ampezzo, 1962; Turin, 1965; Trieste, 1969; Florence, 1972; Siena, 1975; Assisi, 1978; and Como, 1981). These symposia, through their emphasis on informal exchange of ideas and discussions rather than on formal presentations, have proved enormously fruitful for the development of theoretical geodesy. His ideas have greatly influenced scientists such as Martin Hotine and Erik Grafarend.

The roots of Marussi's thinking can be found in his mathematical training and in his observational work as a surveyor and engineer. His work on three-dimensional "differential" or "intrinsic" geodesy was motivated by his desire to use only observable quantities having physical reality, and to use them in a way as direct and unadulterated as possible, avoiding the "epicycles" of reductions so dear to classical geodesy. The courage and firmness with which he had to defend his ideas in the beginning was based on strong philosophical convictions, and the esthetic beauty of his work reveals his artistic background, which we mentioned at the beginning.

Let it not be forgotten, however, that Marussi was not a pale theoretician working in an ivory tower. He was a renaissance personality in the full meaning of the term: a strong, sometimes very strong, character with universal interests, a great mountaineer and traveler, and an able scientific administrator, not lacking political convictions and not afraid of expressing them.

As a geophysicist, he participated in Italian expeditions in the Karakorum and Hindukush mountains in 1954, 1955, and 1961, where he reached an elevation of 6233 m starting from the K2 base camp. These expeditions laid the basis for his love of Eastern culture, art, and philosophy. His scientific status, his personal friendship with scientists from different countries, his diplomatic abilities, and last but not least his tenacious delight in overcoming difficult problems, made it possible to organize a geophysical campaign in the difficult border areas of Soviet Union, India, Pakistan and Afghanistan in 1974, 1975 and 1978: in cooperation with Soviet and Indian geophysicists he established a seismic profile starting from Lake Karakul (USSR) through Pamir, Karakorum and Kashmir to the Indus and Ganges plain.

His love of Eastern culture and people made him one of the first Western scientists to travel to the People's Republic of China (1975) and to Tibet (1981), a fact of which he was particularly proud.

Also at home he loved to perform geodetic and geophysical experimental work. Inspired by the International Geophysical Year (1957 – 1958), he installed a geophysical station in the Grotta Gigante in the karst region near Trieste. There, his group was one of the first to observe, by means of horizontal pendulums, the free oscillations of the earth following the great Chilean earthquake in 1960 (such free oscillations had been predicted theoretically only shortly before).

He also worked on constructing an apparatus for measuring the gravitational constant, and even a satellite observation camera. Only recently he initiated an

Italian-German-Austrian collaboration in the Project ALGEDOP (*Al*pine *Ge*oid by *Dop*pler), in order to strengthen the geoid determination in the Alpine region by Doppler observations.

Marussi's achievements received highest recognition. He became Member of the Accademia Nazionale dei Lincei in 1961, and of the Royal Astronomical Society in London in 1969. He was President of the International Association of Geodesy from 1967 to 1971 and then its Honorary President. For many years he was President of the Italian Geodetic Commission. In 1972 he became Member of the Council of the University of the United Nations. In 1977 he was awarded the Honorary Doctor's Degree in Technical Sciences by the Technical University at Graz, Austria.

A small group of friends visited Antonio Marussi in his home in Trieste on his 75th birthday on October 12, 1983. His always splendid health had failed him about a year before, when a perfidious paralysis confined him to an armchair: refusing a wheelchair, he always insisted on walking, painfully enough. But his mind was absolutely lucid and serene. (The word "olympian" comes to my mind: his leonine appearance reminded me of Goethe.) His wonderful wife, Lori, cared for him and for the guests, and he told us that, 2 days before, he had sent two associates to Nepal to realize his old dream, to clarify the geological relation between the Ganges plain, the Himalaya, and Tibet.

Antonio Marussi passed away gently, without suffering, while working at his desk on April 24, 1984.

Helmut Moritz

Name Index

Subject Index